Progress in Mathematics
Volume 159

Series Editors
Hyman Bass
Joseph Oesterlé
Alan Weinstein

Sundaram Thangavelu

Harmonic Analysis
on the
Heisenberg Group

Birkhäuser
Boston • Basel • Berlin

Sundaram Thangavelu
Indian Statistical Institute
Statistics & Mathematics Division
Bangalore
560 059 India

Library of Congress Cataloging-in-Publication Data

Thangavelu, Sundaram.
    Harmonic analysis on the Heisenberg group / Sundaram Thangavelu.
       p.   cm. -- (Progress in mathematics ; v. 159)
    Includes bibliographical references and index.
    ISBN 0-8176-4050-9 (alk. paper). -- ISBN 3-7643-4050-9 (alk.
paper)
    1. Harmonic analysis. 2. Nilpotent Lie groups. I. Title.
II. Series: Progress in mathematics (Boston, Mass.) ; vol. 159.
QA403.T53   1998   97-47315
515'.2433--dc21   CIP

AMS Codes: 43A30, 43A85, 43A90, 43A55, 43A50, 43A20, 43A15, 43A10,
42B05, 42B08, 42B15, 42B20, 42B25, 22D15, 22D20, 22E27

Printed on acid-free paper
© 1998 Birkhäuser

Birkhäuser

ISBN 0-8176-4050-9
ISBN 3-7643-4050-9

Typeset by the author in TEX
Printed and bound by Quinn Woodbine, Woodbine, NJ

9 8 7 6 5 4 3 2 1

To my family

Scorn not the castle, Architect,
It's nothing but a child's play!
Scorn not the sonnet, Critic,
It's only ignorance on display!

Preface

This monograph deals with various aspects of harmonic analysis on the Heisenberg group. The Heisenberg group is the most well known example from the realm of nilpotent Lie groups and plays an important role in several branches of mathematics, such as representation theory, partial differential equations, several complex variables and number theory. As it is the 'most commutative' among the noncommutative Lie groups, it offers the greatest opportunity for generalising the remarkable results of Euclidean harmonic analysis.

My aim in this work is to demonstrate how standard results of abelian harmonic analysis, such as Plancherel and Paley-Wiener theorems, Wiener-Tauberian theorems, Bochner-Riesz means and multipliers for the Fourier transform, and so on, take shape in the noncommutative setup of the Heisenberg group. Basic results about the representations and the Fourier transform are covered in the first chapter. There are many good texts dealing with these basic results (see Folland [26]) but most of them stop there to develop different topics. Here, however, we pursue a detailed study of the Fourier transform which goes well beyond the basic Stone-von Neumann theorem. We demonstrate the beautiful interplay between the representation theory on the Heisenberg group and the classical expansions in terms of Hermite and Laguerre functions. We prove analogues of Paley-Wiener theorems and Hardy's theorem for the group Fourier transform.

In the second chapter, we develop the spectral theory of the sublaplacian following Strichartz. The eigenfunctions of the sublaplacian are given in terms of the special Hermite functions. There results expansions of functions in terms of these eigenfunctions, sort of a Peter-Weyl theorem for the Heisenberg group. We prove an Abel summability result for these expansions. Then we go on to study the mapping properties of the spectral projections associated to these expansions and prove Müller's restriction theorem. Using this, we study the Bochner-Riesz means associated to the sublaplacian. We also a develop the Littlewood-Paley-Stein theory and prove a weaker version of the multiplier theorem for the sublaplacian.

A study of the group algebra $L^1(H^n/U(n))$ is undertaken in chapter 3 and some applications are given. The Heisenberg group H^n and the unitary group $U(n)$ form a Gelfand pair. We study the elementary spherical functions associated to this pair and prove versions of Wiener-Tauberian theorem. This part of the chapter has some overlap with the work of Faraut-Harzallah [21]. Using the Wiener-Tauberian theorem and the summability result of Strichartz, we study the injectivity of the spherical mean value operator. We also prove a maximal theorem for the spherical means. In the last chapter, we consider the reduced Heisenberg group, and in that context improve some of the theorems treated in previous chapters.

We do not use any major result from the representation theory of Lie groups. However, we use many results from Euclidean harmonic analysis. In fact, as we have already remarked, our aim is to develop several topics from the classical Fourier analysis in the noncommutative setup of the Heisenberg group. The reader is therefore expected to have a good foundation of Euclidean harmonic analysis. We recommend the books *An Introduction to Fourier Analysis on Euclidean Spaces* by E.M. Stein and G. Weiss and [62] of C. Sogge. We use standard notations followed in the above-mentioned books. However, we would like to warn the reader that due to the shortage of new notations, we have used the same symbol to denote the Euclidean Fourier transform as well as the group Fourier transform. Similarly, the Fourier-Weyl transform and the partial Fourier transform are denoted by the same symbol. We hope that the context will make it clear which transform is being considered.

This work is an outgrowth of the lecture notes of a course I gave at UNM, Albuquerque, during the spring of 1997. Earlier in 1994, during the Harmonic Analysis meeting in I.I.T., Mumbai, I gave a series of five lectures on the theme of harmonic analysis on the Heisenberg group. My aim had been to show, without proofs, how the standard results of Euclidean harmonic analysis look in the context of the Heisenberg group. Later, in 1996, I elaborated on some of the topics and gave a series of lectures in the I.S.I winter school held in New Delhi. Ever since I have been contemplating expanding those lectures into a monograph that could serve as a full length text for a course. This past spring, the Department of Mathematics and Statistics gave me comfortable chair to carry out the plan.

I have chosen the topics in this book according to my taste and understanding. To keep the exposition simple, some results are stated without proof. In some cases, I have sacrificed optimality for the sake of simplicity. I have indicated some conjectures and there are many open problems worthy of further investigation. I am afraid that while this work may not describe any great peak in the world of mathematics, there will be some enchanting mesas to be enjoyed. In the desert sand of these pages, I hope, the careful reader will find some wild flowers of cactus and yucca.

It is a great pleasure to express my gratitude to various people who made this monograph possible. First of all, I am grateful to Alladi Sitaram for persuading me to give lectures on the Heisenberg group in the I.S.I winter school. I thank my friends Jay Epperson and Cristina Pereyra and the students C. Dochitoiu and S. Zheng who attended my lectures with enthusiasm. The encouraging remarks of G.B. Folland and R. S. Strichartz are gratefully acknowledged.

It goes without saying that I am immensely thankful to my wife and daughters — my little ϵ and δ — for keeping me relaxed during the preparation of these notes. I am also thankful to all our Indian friends in Albuquerque who made our stay here enjoyable. Finally, I wish to thank the Indian Statistical Institute for giving me leave and the Department of Mathematics and Statistics at UNM for providing me with excellent facilities and warm hospitality.

S. Thangavelu
Albuquerque
May, 1997.

The referees made a careful and thorough study of the manuscript. I have incorporated several modifications in this final version from their suggestions for which I am grateful. I wish to thank M. Sundari for her meticulous proofreading of the manuscript and for preparing the index. It has been a pleasure working with the staff of Birkhäuser in publishing this monograph. Their kind cooperation is thankfully acknowledged.

S. Thangavelu,
Bangalore,
November, 1997.

Contents

Harmonic Analysis
on the
Heisenberg Group

Chapter 1

THE GROUP FOURIER TRANSFORM

In this chapter we introduce the Heisenberg group and study the group Fourier transform. The Heisenberg group is constructed as a group of unitary operators acting on $L^2(\mathbb{R}^n)$. All its irreducible, unitary representations are identified using a theorem of Stone and von Neumann. Then the group Fourier transform is defined and basic results such as the Plancherel theorem and the inversion formula are proved. To further study the properties of the Fourier transform, we introduce the Hermite and special Hermite functions. We prove versions of the Paley-Wiener theorem and Hardy's theorem for the Fourier transform on the Heisenberg group.

1.1 The Heisenberg group

The Heisenberg group plays an important role in several branches of mathematics such as representation theory, harmonic analysis, several complex variables, partial differential equations and quantum mechanics. There are, therefore, several ways of realising the group. What is most remarkable about this group is that it arises in two fundamental but different settings. On the one hand, it can be realised as a group of unitary operators generated by the exponentials of the position and momentum operators in quantum mechanics. On the other hand it can be identified with the group of translations of the Siegel upper half space in $\mathbb{C}^{n+1}$ and plays an important role in our understanding of several problems in the complex function theory of the unit ball.

Classical commutative Fourier analysis on $\mathbb{R}^n$ deals with the Fourier transform $\hat{f} = \mathcal{F}f$ defined by

$$\hat{f}(\xi) = (2\pi)^{-\frac{n}{2}} \int_{\mathbb{R}^n} e^{-ix.\xi} f(x) dx.$$

Considered an operator acting on $L^2(\mathbb{R}^n)$, the Fourier transform $\mathcal{F}$ is a unitary operator. Apart from $\mathcal{F}$ we also have two groups of unitary operators acting on the same Hilbert space. Define $e(x)$ and $\tau(y)$ on $L^2(\mathbb{R}^n)$ by

$$e(x)f(\xi) = e^{ix.\xi} f(\xi), \quad \tau(y)f(\xi) = f(\xi + y).$$

Here x and y are in $\mathbb{R}^n$. Then it is clear that $\{e(x) : x \in \mathbb{R}^n\}$ and $\{\tau(y) : y \in \mathbb{R}^n\}$ are groups of unitary operators and each is isomorphic to $\mathbb{R}^n$. The important fact is that the Fourier transform intertwines these two groups: $\mathcal{F}\tau(y)\mathcal{F}^{-1} = e(y)$. Classical Fourier analysis deals with the interplay between these unitary operators. We also have rotations and dilations coming into the picture.

The operators $e(x)$ and $\tau(y)$ are the unitary operators generated by the position and momentum operators in quantum mechanics. Let Q_j, D_j, $j = 1, 2, \ldots, n$ be the unbounded operators defined on suitable domains by

$$Q_j f(\xi) = \xi_j f(\xi), \quad D_j f(\xi) = -i\frac{\partial}{\partial \xi_j} f(\xi).$$

For every $x, y \in \mathbb{R}^n$ we define

$$x.Q = \sum_{j=1}^n x_j.Q_j, \quad y.D = \sum_{j=1}^n y_j.D_j.$$

Then the operators $ix.Q$ and $iy.D$ are skew Hermitian. By Stone's theorem $\exp(ix.Q)$ and $\exp(iy.D)$ are unitary operators, and we have

$$e(x) = exp(ix.Q), \tau(y) = exp(iy.D).$$

The basic commutation relations between the operators Q_j and D_j are

$$[Q_j, D_j] = iI, j = 1, 2, \ldots, n$$

where I is the identity operator. All other commutators are zero. On the level of the groups $\{e(x) : x \in \mathbb{R}^n\}$ and $\{\tau(y) : y \in \mathbb{R}^n\}$ the commutation relations take the form

$$e(x)\tau(y) = e^{-ix.y}\tau(y)e(x). \qquad (1.1.1)$$

Using (1.1.1) one easily calculates that

$$e(x)\tau(y)e(u)\tau(v) = e^{iy.u}e(x)e(u)\tau(y)\tau(v) = e^{iy.u}e(x+u)\tau(y+v).$$

The above formula shows that the set $\{e(x)\tau(y) : x, y \in \mathbb{R}^n\}$ is not closed under multiplication. We augment this set by including the operators

$$\chi(t)f(\xi) = e^{it}f(\xi), t \in \mathbb{R}$$

and consider the set

$$G = \{e(x)\tau(y)\chi(t) : x, y \in \mathbb{R}^n, t \in \mathbb{R}\}.$$

An easy calculation shows that

$$e(x)\tau(y)\chi(t)e(u)\tau(v)\chi(s) = e(x+u)\tau(y+v)\chi(t+s+u.y),$$

where $u.y$ is the Euclidean inner product. This means that G is a group of unitary operators. Therefore, we can turn $\mathbb{R}^n \times \mathbb{R}^n \times \mathbb{R}$ into a nonabelian group by defining the group operation as

$$(x, y, t)(u, v, s) = (x+u, y+v, t+s+u.y). \qquad (1.1.2)$$

This group is called the polarised Heisenberg group and is denoted by H_{pol}^n.

In the above definition the group operation is not symmetric in all the variables. Instead of considering the exponentials of $ix.Q$ and $iy.D$ separately, we can consider $\pi(x, y) = \exp i(x.Q + y.D)$. It can be verified that

$$\pi(x, y) = e^{\frac{i}{2}x.y}e(x)\tau(y). \qquad (1.1.3)$$

If we set $\pi(x, y, t) = \chi(t)\pi(x, y)$ then the set

$$\{\pi(x, y, t) : x, y \in R^n, t \in R\}$$

becomes a group. In fact, an easy calculation shows that

$$\pi(x, y, t)\pi(u, v, s) = \pi(x+u, y+v, t+s+\frac{1}{2}(u.y - v.x)). \qquad (1.1.4)$$

Hence we can make $\mathbb{R}^n \times \mathbb{R}^n \times \mathbb{R}$ into a group by defining

$$(x, y, t)(u, v, s) = (x+u, y+v, t+s+\frac{1}{2}(u.y - v.x)). \qquad (1.1.5)$$

This group is called the Heisenberg group and is denoted by H^n.

The groups H^n_{pol} and H^n can be realised as groups of upper triangular matrices. To each $(x, y, t) \in H^n_{pol}$ we can associate the $(n + 2) \times (n + 2)$ matrix

$$m(x, y, t) = I + \begin{pmatrix} 0 & y & t \\ 0 & 0 & x \\ 0 & 0 & 0 \end{pmatrix}.$$

It is easy to see that

$$m(x, y, t)m(u, v, s) = m(x + u, y + v, t + s + u.y). \qquad (1.1.6)$$

and hence the set $\{m(x, y, t) : x, y \in \mathbb{R}^n, t \in \mathbb{R}\}$ is a group under matrix multiplication. The polarised Heisenberg group is isomorphic to this group with the isomorphism being given by $(x, y, t) \to m(x, y, t)$. Similarly, we can identify the Heisenberg group with the group of matrices of the form $m(x, y, t + \frac{1}{2}x.y)$.

The set of all matrices of the form $M(x, y, t) = m(x, y, t) - I$ is a Lie algebra with the usual Lie bracket $[A, B] = AB - BA$. We have $M(x, y, t)^2 = M(0, 0, x.y)$ and for $k \geq 3, M(x, y, t)^k = 0$. This means that the Lie algebra $\{M(x, y, t) : x, y \in \mathbb{R}^n, t \in \mathbb{R}\}$ is nilpotent. If $\exp A$ stands for the exponential of the matrix A, then a calculation shows that

$$\exp M(x, y, t) = m(x, y, t + \frac{1}{2}x.y).$$

Hence the Heisenberg group is the image of the above Lie algebra under the exponential map. We can make $\mathbb{R}^n \times \mathbb{R}^n \times \mathbb{R}$ into a Lie algebra by defining a Lie bracket

$$[(x, y, t), (u, v, s)] = (0, 0, u.y - v.x) \qquad (1.1.7)$$

and this Lie algebra is isomorphic to the above Lie algebra of matrices.

In the Heisenberg group we have $(2n+1)$ one parameter subgroups given by

$$G_j = \{(te_j, 0, 0) : t \in \mathbb{R}\}, \quad G_{n+j} = \{(0, te_j, 0) : t \in \mathbb{R}\},$$

$1 \leq j \leq n$ and $G_{2n+1} = \{(0, 0, t) : t \in \mathbb{R}\}$ where e_j are the coordinate vectors in $\mathbb{R}^n$. Corresponding to these one parameter subgroups, we have $(2n + 1)$ left invariant vector fields. These are given by

$$X_j = \left(\frac{\partial}{\partial x_j} - \frac{1}{2}y_j\frac{\partial}{\partial t}\right) \qquad j = 1, 2, \ldots, n,$$

$$Y_j = \left(\frac{\partial}{\partial y_j} + \frac{1}{2} x_j \frac{\partial}{\partial t} \right) \qquad j = 1, 2, \ldots, n,$$

and $T = \frac{\partial}{\partial t}$. The $(2n + 1)$ vector fields generate the Lie algebra h_n of the Heisenberg group. The only nontrivial commutation relations are

$$[X_j, Y_j] = T, \, j = 1, 2, \ldots, n. \qquad (1.1.8)$$

As is well known, the Lie algebra of vector fields h_n is isomorphic to the Lie algebra of matrices $M(x, y, t)$ defined above. Consequently, h_n is a nilpotent Lie algebra and H^n is a nilpotent Lie group.

Identifying $\mathbb{R}^n \times \mathbb{R}^n$ with $\mathbb{C}^n$ the symplectic form $(u.y - v.x)$ can be written as $Im(z.\bar{w})$ where $z = x + iy$ and $w = u + iv$. Thus we can define $H^n = \mathbb{C}^n \times \mathbb{R}$ with the group law given by

$$(z, t)(w, s) = (z + w, t + s + \frac{1}{2} Im(z.\bar{w})). \qquad (1.1.9)$$

For most of this monograph we stick to this complex notation. In the above $z.\bar{w} = z_1 \bar{w}_1 + \ldots + z_n \bar{w}_n$ is the standard Hermitian form on $\mathbb{C}^n$. We have already constructed the Heisenberg group from the basic operators of quantum mechanics. Now we will see how the Heisenberg group arises in several complex variables.

It is well known that the unit disk $|w| < 1$ in the complex plane can be mapped onto the upper half plane $Im(z) > 0$ by a fractional linear transformation which is explicitly given by

$$w = \left(\frac{i - z}{i + z} \right), \, z = i \left(\frac{1 - w}{1 + w} \right).$$

The group $\mathbb{R}$ of real numbers acts on the upper half plane by horizontal translations $z \to z + t, t \in \mathbb{R}$. Thus we can identify $\mathbb{R}$ with the boundary of the upper half plane and this identification leads to many important consequences. Many basic operators, such as the Cauchy integral, can be identified with convolution operators on $\mathbb{R}$. There is a similar situation in the higher dimensional case as well.

Let B_{n+1} denote the unit ball $|w| < 1$ in $\mathbb{C}^{n+1}$ and let S_{n+1} be the Siegel upper half space defined by

$$S_{n+1} = \{(z, z_{n+1}) \in \mathbb{C}^{n+1} : Im(z_{n+1}) > |z|^2 \}. \qquad (1.1.10)$$

The two domains are biholomorphically equivalent. Indeed, the fractional linear transformation $\varphi(w, w_{n+1}) = (\frac{w}{1+w_{n+1}}, i \frac{1-w_{n+1}}{1+w_{n+1}})$ maps B_{n+1}

onto S_{n+1}. This transformation also preserves boundaries: φ takes the boundary of the unit ball onto the boundary ∂S_{n+1} which is given by $Im(z_{n+1}) = |z|^2$. The domain S_{n+1} is invariant under a large group of symmetries: dilations, rotations and translations. For each $r > 0$, if we define

$$\delta_r(z, z_{n+1}) = (rz, r^2 z_{n+1}) \qquad (1.1.11)$$

then it is clear that S_{n+1} is invariant under the action of δ_r. This non-isotropic dilations play an important role in the analysis of convolution operators on the Heisenberg group. For each unitary map σ of $\mathbb{C}^n$, we have the rotation $(z, z_{n+1}) \rightarrow (\sigma z, z_{n+1})$ and it is clear that S_{n+1} is also invariant under these maps. Finally the translations of S_{n+1} are given by the Heisenberg group; this enables us to identify H^n with the boundary ∂S_{n+1} of the Siegel upper half space.

For every $(\zeta, t) \in \mathbb{C}^n \times \mathbb{R}$, consider the transformation $L(\zeta, t)$ acting on S_{n+1} by

$$L(\zeta, t)(z, z_{n+1}) = (z + \zeta, z_{n+1} + t + 2iIm(z.\bar{\zeta}) + i|\zeta|^2). \quad (1.1.12)$$

The map $L(\zeta, t)$ takes S_{n+1} into itself since

$$|z + \zeta|^2 - |z|^2 = Im(2iz.\bar{\zeta} + i|\zeta|^2). \qquad (1.1.13)$$

It also preserves the boundary. If we compose $L(\zeta, t)$ with $L(\zeta', t')$ we obtain

$$L(\zeta, t)L(\zeta', t') = L(\zeta + \bar{\zeta}', t + t' + \frac{1}{2}Im(\zeta.\bar{\zeta}')).$$

This follows from the identity

$$2i\zeta.\bar{\zeta}' + i|\zeta|^2 + i|\zeta'|^2 = 2Im(\zeta.\bar{\zeta}') + i|\zeta + \zeta'|^2.$$

Thus (1.1.13) gives a realisation of the H^n as a group of affine holomorphic bijections of S_{n+1}.

It is important to note that the mappings (1.1.13) are simply transitive on the boundary: given any two points on the boundary there is exactly one element of the Heisenberg group mapping the first to the second. In particular

$$L(\zeta, t)(0, 0) = (\zeta, t + i|\zeta|^2)$$

and hence we can identify the Heisenberg group with the boundary ∂S_{n+1} via the action on the origin. So the Heisenberg group acts on itself and the action is nothing but the left translation on the Heisenberg group.

Since the action of $L(\zeta, t)$ on H^n is holomorphic, the Cauchy-Riemann operators on S_{n+1} are invariant under it. Hence they can be realised as convolution operators on the Heisenberg group and Fourier analytic techniques can be used to study them in detail. This point of view has been proved to be very useful in the study of some problems on the Siegel upper half space. For more about the role of the Heisenberg group in several complex variables, we refer the reader to Stein [65] and Folland and Kohn [24].

1.2   The Schrödinger representations

The representation theory of the Heisenberg group is fairly simple and well understood. Using a fundamental theorem called the Stone-von Neumann theorem, we can give a complete classification of all the irreducible unitary representations of the Heisenberg group.

First we recall a few definitions from representation theory. Let G be a group and $\mathcal{H}$ a Hilbert space. Denote by $\mathcal{U}(\mathcal{H})$ the group of unitary operators acting on $\mathcal{H}$. A homomorphism π of G into $\mathcal{U}(\mathcal{H})$ is said to be a strongly continuous unitary representation of G on $\mathcal{H}$, if for every $x \in \mathcal{H}$, the map $g \rightarrow \pi(g)x$ is continuous. A subspace M of $\mathcal{H}$ is said to be invariant under π if $\pi(g)x$ belongs to M for all g in G whenever x is in M. A unitary representation π is said to be irreducible if there is no nontrivial proper closed subspace of M that is invariant under π. Two representations π and ρ are said to be unitarily equivalent if there is $T \in \mathcal{U}(\mathcal{H})$ such that $\rho(g) = T\pi(g)T^*$ for all g.

With these definitions, we now turn our attention to the representations of the polarised Heisenberg group. From the definition of the group operation, it follows that the map

$$(x, y, t) \rightarrow \rho(x, y, t) = \chi(t)e(x)\tau(y)$$

defines a unitary representation of the polarised Heisenberg group. This representation is irreducible. The following arguments convince us that this is indeed the case (we will see another proof shortly). Suppose an operator commutes with all $e(x)$. Then it is given by multiplication by a bounded function. This can be proved by expanding compactly supported functions into a Fourier series. If the operator also commutes with $\tau(y)$, then it has to be a scalar multiple of the identity. Hence if

A commutes with all $\rho(x, y, t)$ then it has to be a scalar multiple of the identity and, by Schur's lemma, ρ is irreducible.

The above arguments depend heavily on results from representation theory. We will now give a direct proof using the Plancherel theorem for the Euclidean Fourier transform. We will show that ρ is irreducible by establishing that a related representation of the Heisenberg group is irreducible. Let $\pi(x, y, t)$ be defined by

$$\pi(x, y, t) = e^{it} e^{\frac{i}{2} x \cdot y} e(x) \tau(y).$$

Then, from the definition of the Heisenberg group, it follows that $(x, y, t) \to \pi(x, y, t)$ is a unitary representation of the Heisenberg group. We will show by using the Euclidean Plancherel theorem that π is indeed irreducible. More generally, for every $\lambda \neq 0$ define $\pi_\lambda(x, y, t) = \pi(\lambda x, y, \lambda t)$. Thus

$$\pi_\lambda(x, y, t)\varphi(\xi) = e^{i\lambda t} e^{i\lambda(x \cdot \xi + \frac{1}{2} x \cdot y)} \varphi(\xi + y) \qquad (1.2.1)$$

where $\varphi \in L^2(\mathbb{R}^n)$. A celebrated theorem of Stone-von Neumann says that up to unitary equivalence these are all the irreducible unitary representations of H^n that are nontrivial at the center.

Theorem 1.2.1 *The representations $\pi_\lambda, \lambda \neq 0$ are irreducible. If ρ is an irreducible unitary representation of H^n on a Hilbert space $\mathcal{H}$ such that $\rho(0, 0, t) = e^{i\lambda t} I$ for some $\lambda \neq 0$, then ρ is unitarily equivalent to π_λ.*

For a proof of the Stone-von Neumann theorem, we refer to Folland [26]. The proof uses many properties of the Fourier-Wigner and Weyl transforms. Here we will show that π_λ are irreducible. We do this when $\lambda = 1$; the general case is similar. In order to prove this we use the Plancherel theorem for the Euclidean Fourier transform. Let us write $\pi_1(x, y, t) = e^{it} \pi(z)$ where $z = x + iy$ and

$$\pi(z)\varphi(\xi) = e^{i(x \cdot \xi + \frac{1}{2} x \cdot y)} \varphi(\xi + y).$$

Suppose $M \subset L^2(\mathbb{R}^n)$ is invariant under all $\pi_1(x, y, t)$. If $M \neq \{0\}$ we will show that $M = L^2(\mathbb{R}^n)$ proving the irreducibility of π_1.

If M is a proper subspace of $L^2(\mathbb{R}^n)$ invariant under $\pi_1(x, y, t)$ for all (x, y, t), then there are nontrivial functions f and g in $L^2(\mathbb{R}^n)$ such that $f \in M$ and g is orthogonal to $\pi(z)f$ for all z. This means that $(\pi(z)f, g) = 0$ for all z.

Now, given $\varphi, \psi \in L^2(\mathbb{R}^n)$, consider the function

$$V_\varphi(\psi, z) = (2\pi)^{-\frac{n}{2}} (\pi(z)\varphi, \psi).$$

This is called the Fourier-Wigner transform of φ and ψ. Let us calculate the $L^2(\mathbb{C}^n)$ norm of this function. Explicitly,

$$V_\varphi(\psi, z) = (2\pi)^{-\frac{n}{2}} \int_{\mathbb{R}^n} e^{ix.\xi} \varphi(\xi + \frac{y}{2}) \bar{\psi}(\xi - \frac{y}{2}) \, d\xi. \qquad (1.2.2)$$

Applying the Plancherel theorem for the Fourier transform in the x variable, we get

$$\int_{\mathbb{C}^n} |V_\varphi(\psi, z)|^2 \, dz = \int_{\mathbb{R}^{2n}} |\varphi(\xi + \frac{y}{2})|^2 |\bar{\psi}(\xi - \frac{y}{2})|^2 \, d\xi dy \qquad (1.2.3)$$

which after making a change of variables becomes

$$\int_{\mathbb{R}^n} |\varphi(\xi)|^2 \, d\xi \int_{\mathbb{R}^n} |\bar{\psi}(\xi)|^2 \, d\xi. \qquad (1.2.4)$$

Thus we have $\|V_\varphi(\psi)\|_2 = \|\varphi\|_2 \|\psi\|_2$. Under our assumption that M is nontrivial and proper, we have $V_f(g) = 0$ which means $\|f\|_2 \|g\|_2 = 0$; this is a contradiction since both f and g are nontrivial. Hence M has to be the whole of $L^2(\mathbb{R}^n)$ and this proves that π_1 is irreducible. ∎

The above property of the Fourier -Wigner transform is very useful in the study of the group Fourier transform. Let us note this in the following proposition.

Proposition 1.2.2 *Let* φ, ψ, f, g *be in* $L^2(\mathbb{R}^n)$. *Then*

$$(V_\varphi(\psi), V_f(g)) = (\varphi, f)(g, \psi).$$

The proposition follows from (1.2.3) by polarisation. The Fourier-Wigner transform and its close relative, the Wigner transform, have many interesting properties. See Folland [26] for more details.

As we have seen above, $V_\varphi(\psi)$ is in $L^2(\mathbb{C}^n)$. By Hölder's inequality and the Riemann-Lebesgue lemma for the Fourier transform, we also know that $V_\varphi(\psi)$ is a bounded continuous function vanishing at infinity. An interpolation gives the following.

Corollary 1.2.3 *For* $\varphi, \psi \in L^2(\mathbb{R}^n)$, $V_\varphi(\psi) \in L^p(\mathbb{C}^n)$ *for* $2 \leq p \leq \infty$.

The corollary will be used in defining the Fourier transform of L^p functions for $1 \leq p \leq 2$. We note that the proposition has the following important consequence. Suppose $\{\varphi_j\}$ is an orthonormal system in $L^2(\mathbb{R}^n)$ and define Φ_{jk} to be the Fourier-Wigner transform of φ_j and φ_k. Then from Proposition 1.2.2, it follows that $\{\Phi_{jk}\}$ is an orthonormal system in $L^2(\mathbb{C}^n)$. Actually, we will show that $\{\Phi_{jk}\}$ is an orthonormal basis whenever $\{\varphi_j\}$ is. We will prove this after examining the Weyl transform.

We conclude this section by noting the following consequence of the Stone-von Neumann theorem.

Theorem 1.2.4 *Every irreducible unitary representation of H^n is unitarily equivalent to one and only one of the following representations:(i) $\pi_\lambda(x,y,t), \lambda \in \mathbb{R}^*$, acting on $L^2(\mathbb{R}^n)$; (ii) $\chi_{(\xi+i\eta)}(x,y,t) = e^{i(x.\xi+y.\eta)}$, acting on $\mathbb{C}$.*

Proof: Let $Z = \{(0,0,t) : t \in \mathbb{R}\}$ be the centre of the Heisenberg group. If ρ is an irreducible unitary representation, then by Schur's lemma, ρ must map Z homomorphically into the group $\{cI : |c| = 1\}$ so that $\rho(0,0,t) = e^{i\lambda t}I$ for some λ real. If λ is nonzero, then by the theorem of Stone and von Neumann, ρ is unitarily equivalent to π_λ. If $\lambda = 0$, then ρ factors through the quotient group H^n/Z which is isomorphic to $\mathbb{C}^n$. The irreducible unitary representations of $\mathbb{C}^n$ are all one-dimensional; they are the homomorphisms

$$(x,y) \rightarrow e^{i(x.\xi+y.\eta)}$$

into the circle group. Hence ρ will be equivalent to some $\chi_{(\xi+i\eta)}$. This proves the theorem. ∎

In this section we have considered only the Schrödinger picture. That is, all the representations we considered are realised on the Hilbert space $L^2(\mathbb{R}^n)$. There is another picture, known as the Fock space realisation, in which the representations act on certain spaces of entire functions. We refer to Folland [26] for details.

1.3 The Fourier and Weyl transforms

Using the representations of the Heisenberg group described in the previous section, we now define the group Fourier transform on the Heisenberg group as an operator valued function. For our definition we use the Schrödinger picture so that, for a function f on H^n, the associated Fourier transform will be a family of operators acting on the same Hilbert space $L^2(\mathbb{R}^n)$.

Since the Heisenberg group H^n equals $\mathbb{C}^n \times \mathbb{R}$ as a set we have the Lebesgue measure $dzdt$ on H^n. It is easily seen that this measure is both right and left translation invariant. Hence it is the Haar measure on H^n and the group is unimodular. With this measure, we form the usual function spaces $L^p(H^n)$.

First we define the Fourier transform for integrable functions f. For each $\lambda \in \mathbb{R}^*$, $\hat{f}(\lambda)$ is the operator acting on $L^2(\mathbb{R}^n)$ given by

$$\hat{f}(\lambda)\varphi = \int_{H^n} f(z,t)\pi_\lambda(z,t)\varphi \, dzdt. \qquad (1.3.5)$$

Here $\pi_\lambda(z,t)$ are the Schrödinger representations and the integral is a Bochner integral taking values in the Hilbert space $L^2(\mathbb{R}^n)$. If ψ is another function in $L^2(\mathbb{R}^n)$, then

$$(\hat{f}(\lambda)\varphi, \psi) = \int_{H^n} f(z,t)(\pi_\lambda(z,t)\varphi, \psi) \, dzdt.$$

Since $\pi_\lambda(z,t)$ are unitary operators, it follows that

$$|(\pi_\lambda(z,t)\varphi, \psi)| \leq \|\varphi\|_2 \|\psi\|_2$$

and consequently

$$|(\hat{f}(\lambda)\varphi, \psi)| \leq \|\varphi\|_2 \|\psi\|_2 \|f\|_1.$$

This shows that $\hat{f}(\lambda)$ is a bounded operator on $L^2(\mathbb{R}^n)$ and the operator norm satisfies $\|\hat{f}(\lambda)\| \leq \|f\|_1$. We will show that when f is also in $L^2(H^n)$, $\hat{f}(\lambda)$ is a Hilbert-Schmidt operator.

Let us write $\pi_\lambda(z,t) = e^{i\lambda t}\pi_\lambda(z)$ where $\pi_\lambda(z) = \pi_\lambda(z,0)$ and define

$$f^\lambda(z) = \int_{-\infty}^{\infty} e^{i\lambda t} f(z,t) \, dt \qquad (1.3.6)$$

to be the inverse Fourier transform of f in the t variable. Then it follows that

$$\hat{f}(\lambda)\varphi = \int_{\mathbb{C}^n} f^\lambda(z)\pi_\lambda(z)\varphi \, dz. \qquad (1.3.7)$$

Thus we are led to consider operators of the form

$$W_\lambda(g) = \int_{\mathbb{C}^n} g(z)\pi_\lambda(z) \, dz$$

for functions on $\mathbb{C}^n$. When $\lambda = 1$, we call this the Weyl transform and denote it by $W(g)$. We also write $\pi(z)$ in place of $\pi_1(z)$. Thus

$$W(g)\varphi(\xi) = \int_{\mathbb{C}^n} g(z)\pi(z)\varphi(\xi)\, dz. \qquad (1.3.8)$$

From the explicit description of the representations, it follows that

$$W(g)\varphi(\xi) = \int_{\mathbb{R}^{2n}} g(x+iy)e^{i(x\cdot\xi+\frac{1}{2}x\cdot y)}\varphi(\xi+y)\, dxdy.$$

Thus $W(g)$ is an integral operator with kernel $K_g(\xi,\eta)$ given by

$$\int_{\mathbb{R}^n} g(x,\eta-\xi)e^{\frac{i}{2}x\cdot(\xi+\eta)}\, dx \qquad (1.3.9)$$

where $g(x,y)$ stands for $g(x+iy)$. Therefore, if $g \in L^1 \cap L^2(\mathbb{C}^n)$, the kernel $K_g(\xi,\eta)$ belongs to $L^2(\mathbb{R}^{2n})$, and hence from the theory of integral operators, it follows that $W(g)$ is a Hilbert-Schmidt operator whose norm is given by

$$\|W(g)\|_{HS}^2 = \int_{\mathbb{R}^{2n}} |K_g(\xi,\eta)|^2\, d\xi d\eta.$$

Using the explicit formula for the kernel and the Plancherel theorem for the Fourier transform, we get

$$\|W(g)\|_{HS}^2 = (2\pi)^n \int_{\mathbb{R}^{2n}} |g(x,y)|^2\, dxdy.$$

This is the Plancherel theorem for the Weyl transform.

We can now establish the Plancherel theorem for the group Fourier transform. Let S_2 denote the Hilbert space of Hilbert-Schmidt operators on $L^2(\mathbb{R}^n)$ with the inner product $(T,S) = tr(TS^*)$. Let $d\mu(\lambda) = (2\pi)^{-n-1}|\lambda|^n d\lambda$ and let $L^2(\mathbb{R}^*, S_2; d\mu)$ stand for the space of functions on $\mathbb{R}^*$ taking values in S_2 and square integrable with respect to $d\mu$.

Theorem 1.3.1 *The group Fourier transform is an isometric isomorphism between $L^2(H^n)$ and $L^2(\mathbb{R}^*, S_2; d\mu)$.*

Proof: First assume that $f \in L^1 \cap L^2(H^n)$. From (1.3.7) it is clear that $\hat{f}(\lambda)$ is an integral operator whose kernel can be explicitly calculated. As in the case of $\lambda = 1$, we can show that

$$|\lambda|^{-n}\|\hat{f}(\lambda)\|_{HS}^2 = (2\pi)^n \int_{\mathbb{C}^n} |f^\lambda(z)|^2\, dz.$$

Recalling the definition of f^λ and integrating with respect to $d\mu$, we get

$$\int \|\hat{f}(\lambda)\|_{HS}^2 \, d\mu(\lambda) = (2\pi)^{-1} \int \int_{\mathbb{C}^n} |f^\lambda(z)|^2 \, dz d\lambda.$$

In view of the Euclidean Plancherel theorem we get

$$\|\hat{f}\|_{L^2(\mathbb{R}^*, S_2; d\mu)} = \|f\|_{L^2(H^n)}.$$

As $L^1 \cap L^2(H^n)$ is dense in $L^2(H^n)$, it is clear from the equality of norms that we can extend the definition of the Fourier transform to all $f \in L^2(H^n)$; the Fourier transform thus defined will verify the above equality of norms. This completes the proof of the theorem. ∎

We also have an inversion theorem for the group Fourier transform. As in the case of the Plancherel theorem, we first consider the Weyl transform. For $w = u + iv \in \mathbb{C}^n$ we have

$$\pi(w)^* W(g) = \int_{\mathbb{C}^n} g(z) \pi(-w) \pi(z) \, dz.$$

But $\pi(-w)\pi(z) = e^{-\frac{i}{2} Im(w.\bar{z})} \pi(z - w)$ so that

$$\pi(w)^* W(g) = \int_{\mathbb{C}^n} g(z) e^{-\frac{i}{2} Im(w.\bar{z})} \pi(z - w) \, dz \qquad (1.3.10)$$

which after a change of variables gives

$$\pi(w)^* W(g) = \int_{\mathbb{C}^n} g(z + w) e^{-\frac{i}{2} Im(w.\bar{z})} \pi(z) \, dz.$$

The above formula shows that $\pi(w)^* W(g) = W(g_w)$ where

$$g_w(z) = g(z + w) e^{-\frac{i}{2} Im(w.\bar{z})}.$$

Then the kernel K_w of $W(g_w)$ can be calculated. We can easily show that

$$K_w(\xi, \xi) = e^{\frac{i}{2} u.v} e^{-iu.\xi} \int_{\mathbb{R}^n} e^{ix.(\xi - \frac{v}{2})} g(x, v) \, dx. \qquad (1.3.11)$$

Now the formula

$$tr(\pi(w)^* W(g)) = \int_{\mathbb{R}^n} K_w(\xi, \xi) \, d\xi$$

gives us

$$tr(\pi(w)^*W(g)) = e^{\frac{i}{2}u.v} \int_{\mathbb{R}^n} e^{-iu.\xi} \Big(\int_{\mathbb{R}^n} e^{ix.(\xi - \frac{v}{2})} g(x,v) \, dx \Big) \, d\xi.$$

In view of the Euclidean inversion formula this leads to

$$tr(\pi(w)^*W(g)) = (2\pi)^n g(w).$$

This is the inversion formula for the Weyl transform. We can now state the inversion formula for the group Fourier transform.

Theorem 1.3.2 *For all Schwartz class functions on H^n, the following inversion formula holds:*

$$f(z,t) = \int_{-\infty}^{\infty} tr(\pi_\lambda^*(z,t)\hat{f}(\lambda)) \, d\mu(\lambda).$$

Proof: For the operators W_λ, one gets an inversion formula similar to the one obtained for the Weyl transform. This follows by a simple change of variables. Together with the inversion theorem for the Fourier transform in the t variable, this then completes the proof. The details are left to the reader. ∎

In the case of the Euclidean Fourier transform, besides the inversion and Plancherel theorems, we also have the Hausdorff-Young inequality. This says that if $f \in L^1 \cap L^p(\mathbb{R}^n), 1 \leq p \leq 2$, then $\hat{f} \in L^q(\mathbb{R}^n)$ where q is the index conjugate to p, and we have $\|\hat{f}\|_q \leq \|f\|_p$. This follows from interpolating the Plancherel theorem with the trivial inequality $\|\hat{f}\|_\infty \leq \|f\|_1$. In the case of the Fourier transform on the Heisenberg group, we have a similar result.

In order to state the result, we need to introduce more notation. For $1 \leq p \leq \infty$, let S_p be the class of compact operators acting on $L^2(\mathbb{R}^n)$ whose singular numbers belong to l^p. In particular, S_1 is the ideal of trace class operators. Each S_p is a Banach space with the norm $\|T\|_p^p = tr(TT^*)^{\frac{p}{2}}$ when p is finite. When $p = \infty$, the norm is taken to be the l^∞ norm of the singular numbers. For more properties of this class of operators, we refer to Reed-Simon [57].

Let $L^p(\mathbb{R}^*, S_p; d\mu(\lambda))$ be the class of functions on $\mathbb{R}^*$ taking values in S_p whose pth powers are integrable with respect to $d\mu(\lambda)$. We now state the Hausdorff-Young theorem for the group Fourier transform.

Theorem 1.3.3 *Let* $1 \leq p \leq 2$. *Then the Fourier transform maps* $L^p(H^n)$ *continuously into* $L^q(R^*, S_q; d\mu(\lambda))$ *and we have the inequality*

$$\|\hat{f}(\lambda)\|_{L^q(R^*, S_q; d\mu(\lambda))} \leq C\|f\|_p$$

where q *is the index conjugate to* p.

The proof of this theorem requires the concepts of noncommutative integration and interpolation. We will not go into the proof, but we refer the reader to the papers of Kunze [37] and Peetre-Sparr [52]. There is a similar inequality for the inverse Fourier transform which will be used later in getting a pointwise estimate for the Riesz kernel associated to the sublaplacian.

We conclude this section with the following remarks. As in the case of $\mathbb{R}^n$, the group Fourier transform takes convolution into products. If f and g are functions on H^n, then their convolution is defined by

$$f * g(z, t) = \int_{H^n} f((z, t)(-w, -s))g(w, s)\, dw\, ds. \qquad (1.3.12)$$

Then it follows from the definition of the Fourier transform that

$$(f * g)\hat{}(\lambda) = \hat{f}(\lambda)\hat{g}(\lambda). \qquad (1.3.13)$$

Convolution of two integrable functions is again integrable. This makes $L^1(H^n)$ into a (noncommutative) Banach algebra. More generally, one has the generalised Young inequality

$$\|f * g\|_r \leq \|f\|_p \|g\|_q$$

where $\frac{1}{r} = \frac{1}{p} + \frac{1}{q} - 1$. This can be proved by an application of the Riesz-Thorin convexity theorem.

Recall the definition of the group Fourier transform:

$$\hat{f}(\lambda) = \int_{\mathbb{C}^n} f^\lambda(z)\pi_\lambda(z)\, dz.$$

From this we observe that as far as the t variable is concerned, the group Fourier transform is nothing but the Euclidean Fourier transform. In many problems on the Heisenberg group, an important technique is to take the partial Fourier transform in the t variable to reduce matters to the case of $\mathbb{C}^n$. Nothing could explain this better than convolution operators on H^n.

Suppose we want to show that the convolution operator taking f into $G * f$ is bounded on $L^2(H^n)$. Then it is clearly enough to show that

$$\int_{\mathbb{C}^n} |(G * f)^\lambda(z)|^2 \, dz \leq C \int_{\mathbb{C}^n} |f^\lambda(z)|^2 \, dz$$

where the constant C is independent of λ. Recalling the definition of the convolution, an easy calculation shows that

$$(G * f)^\lambda(z) = \int_{\mathbb{C}^n} G^\lambda(z - w) f^\lambda(w) e^{-\frac{i}{2}\lambda Im(z.\bar{w})} \, dz.$$

Thus we are led to convolutions of the form

$$g *_\lambda h(z) = \int_{\mathbb{C}^n} g(z - w) h(w) e^{\frac{i}{2}\lambda Im(z.\bar{w})} \, dw. \qquad (1.3.14)$$

These are called the λ-twisted convolutions. When $\lambda = 1$, we simply call them twisted convolutions and denote them by $g \times h$. Thus

$$g \times h(z) = \int_{\mathbb{C}^n} g(z - w) h(w) e^{\frac{i}{2} Im(z.\bar{w})} \, dw. \qquad (1.3.15)$$

The twisted convolution turns $L^1(\mathbb{C}^n)$ into a noncommutative Banach algebra.

It is interesting to note that the Weyl transform bears the same relation to the twisted convolution as the group Fourier transform bears to the convolution. The Weyl transform of $g \times h(z)$ is the product $W(g)W(h)$. This can be seen as follows. Let $\varphi, \psi \in L^2(\mathbb{R}^n)$. Then

$$(W(g)W(h)\varphi, \psi) = \int_{\mathbb{C}^n} g(z)(\pi(z)W(h)\varphi, \psi) \, dz.$$

Writing the definition of $W(h)$ as above, and using the relation $\pi(z)\pi(w) = \pi(z + w)e^{\frac{i}{2}Im(z.\bar{w})}$, we get

$$(W(g)W(h)\varphi, \psi)$$

$$= \int_{\mathbb{C}^n} \int_{\mathbb{C}^n} g(z)h(w)e^{\frac{i}{2}Im(z.\bar{w})}(\pi(z + w)\varphi, \psi) \, dw \, dz.$$

After a change of variable, this shows that

$$(W(g)W(h)\varphi, \psi)$$

$$= \int_{\mathbb{C}^n} \int_{\mathbb{C}^n} g(z - w)h(w)e^{\frac{i}{2}Im(z.\bar{w})}(\pi(z)\varphi, \psi) \, dw \, dz.$$

This proves our claim.

Another interesting property of the twisted convolution which is not shared by the ordinary convolution is the following.

Proposition 1.3.4 *Let $f, g \in L^2(\mathbb{C}^n)$. Then $f *_\lambda g$ also belongs to $L^2(\mathbb{C}^n)$ and we have*

$$\|f *_\lambda g\|_2 \leq C_\lambda \|f\|_2 \|g\|_2.$$

Proof: It is enough to prove this when $\lambda = 1$. The general case follows by dilation. By the Plancherel theorem for the Weyl transform, we have

$$\|f \times g\|_2 = (2\pi)^{-\frac{n}{2}} \|W(f \times g)\|_{HS} = (2\pi)^{-\frac{n}{2}} \|W(f)W(g)\|_{HS}.$$

But $\|TS\|_{HS} \leq \|T\|_{HS}\|S\|_{HS}$ and hence we get the inequality

$$\|f \times g\|_2 \leq (2\pi)^{\frac{n}{2}} \|f\|_2 \|g\|_2,$$

and this proves the proposition. ∎

We remark that the above proposition fails in the case of ordinary convolution. By interpolation we can get many such inequalities for the twisted convolution.

1.4 Hermite and special Hermite functions

In this section we introduce and study some properties of the Hermite and special Hermite functions. These functions are eigenfunctions of the Hermite and special Hermite operators, respectively. The Hermite operator is often called the harmonic oscillator and the special Hermite operator is sometimes called the twisted Laplacian. As we will later see, the two operators are directly related to the sublaplacian on the Heisenberg group. The theory of Hermite and special Hermite expansions is intimately connected to the harmonic analysis on the Heisenberg group. They play an important role in our understanding of several problems on H^n.

We start with the definition of the Hermite polynomials. For $k = 0, 1, 2, \ldots$, and $t \in \mathbb{R}$ we define $H_k(t)$ by the equation

$$H_k(t) = (-1)^k \left(\frac{d^k}{dt^k} \{e^{-t^2}\} e^{t^2} \right). \qquad (1.4.16)$$

The normalised Hermite functions are then defined by

$$h_k(t) = (2^k \sqrt{\pi} k!)^{-\frac{1}{2}} H_k(t) e^{-\frac{1}{2}t^2}. \qquad (1.4.17)$$

These functions form an orthonormal basis for $L^2(\mathbb{R})$. The higher dimensional Hermite functions denoted by Φ_α are then obtained by taking tensor products. Thus for any multi-index α and $x \in \mathbb{R}^n$, we define

$$\Phi_\alpha(x) = \Pi_{j=1}^n h_{\alpha_j}(x_j). \qquad (1.4.18)$$

The family $\{\Phi_\alpha\}$ is then an orthonormal basis for $L^2(\mathbb{R}^n)$. They are eigenfunctions of the Hermite operator $H = -\Delta + |x|^2 : H\Phi_\alpha = (2|\alpha| + n)\Phi_\alpha$ where $|\alpha| = \sum_{j=1}^n \alpha_j$. They are also eigenfunctions of the Fourier transform : $\mathcal{F}\Phi_\alpha = (-i)^{|\alpha|}\Phi_\alpha$. See [84] for a proof. This fact has played a crucial role in some problems in Fourier analysis. For example, Wiener [92] used this in his development of the the Plancherel theorem and Beckner [6] used this property in obtaining the best constant for the Hausdorff-Young inequality.

For each $\alpha, \beta \in \mathbb{N}^n$ and $z \in \mathbb{C}^n$, we define the special Hermite functions $\Phi_{\alpha,\beta}$ by

$$\Phi_{\alpha,\beta}(z) = (2\pi)^{-\frac{n}{2}} \int_{\mathbb{R}^n} e^{ix\cdot\xi} \Phi_\alpha(\xi + \frac{y}{2})\Phi_\beta(\xi - \frac{y}{2}) \, d\xi. \qquad (1.4.19)$$

Thus $\Phi_{\alpha,\beta}(z)$ are the Fourier-Wigner transforms of the Hermite functions Φ_α and Φ_β. We now show that $\{\Phi_{\alpha,\beta}\}$ is an orthonormal basis for $L^2(\mathbb{C}^n)$.

Proposition 1.4.1 *The special Hermite functions form a complete orthonormal system for $L^2(\mathbb{C}^n)$.*

Proof: The orthonormality follows from the properties of the Fourier-Wigner transform. To prove completeness, we use the Plancherel theorem for the Weyl transform. Suppose $f \in L^2(\mathbb{C}^n)$ is orthogonal to all $\Phi_{\alpha,\beta}$. Using the definition of $\Phi_{\alpha,\beta}$ this means that

$$\int_{\mathbb{C}^n} \bar{f}(z)(\pi(z)\Phi_\alpha, \Phi_\beta) \, dz = (W(\bar{f})\Phi_\alpha, \Phi_\beta) = 0.$$

The completeness of $\{\Phi_\alpha\}$ in $L^2(\mathbb{R}^n)$ proves that $W(\bar{f}) = 0$ which implies $f = 0$ in view of the Plancherel theorem for the Weyl transform. ∎

We now state without proof more properties of the special Hermite functions which will be used in later sections (proofs can be found in [84]). The special Hermite functions can be expressed in terms of Laguerre functions. For example, we have the formula

$$\Phi_{\alpha,\alpha}(z) = (2\pi)^{-\frac{n}{2}} \Pi_{j=1}^n L_{\alpha_j}(\frac{1}{2}|z_j|^2)e^{-\frac{1}{4}|z_j|^2} \qquad (1.4.20)$$

where L_k are Laguerre polynomials of type 0. Recall that Laguerre polynomials of type $\delta > -1$ are defined by

$$L_k^\delta(t)e^{-t}t^\delta = \frac{1}{k!}(\frac{d}{dt})^k(e^{-t}t^{k+\delta}). \qquad (1.4.21)$$

More generally, we have the following formulae. For each multi-index m let us define

$$L_\mu^m(z) = \Pi_{j=1}^n L_{\mu_j}^{m_j}(\frac{1}{2}|z_j|^2). \qquad (1.4.22)$$

We then have

Proposition 1.4.2

$$\Phi_{\mu+m,\mu} = (2\pi)^{-\frac{n}{2}}(\frac{\mu!}{(\mu+m)!})^{\frac{1}{2}}(\frac{i}{\sqrt{2}})^{|m|}\bar{z}^m L_\mu^m(z)e^{-\frac{1}{4}|z|^2},$$

$$\Phi_{\mu,\mu+m} = (2\pi)^{-\frac{n}{2}}(\frac{\mu!}{(\mu+m)!})^{\frac{1}{2}}(\frac{-i}{\sqrt{2}})^{|m|}z^m L_\mu^m(z)e^{-\frac{1}{4}|z|^2}.$$

The special Hermite functions are eigenfunctions of an elliptic differential operator on $\mathbb{C}^n$. Indeed, let

$$L = -\Delta_z + \frac{1}{4}|z|^2 - i\sum_{j=1}^n(x_j\frac{\partial}{\partial y_j} - y_j\frac{\partial}{\partial x_j}) \qquad (1.4.23)$$

where Δ_z is the standard Laplacian on $\mathbb{C}^n$. Then one can show that

$$L(\Phi_{\alpha,\beta}) = (2|\beta| + n)\Phi_{\alpha,\beta}.$$

As we will see later, this operator is related to the sublaplacian on the Heisenberg group. We observe that the eigenvalue depends only on β. This means that the $k - th$ eigenspace corresponding to the eigenvalue $(2k + n)$ is infinite dimensional and spanned by $\{\Phi_{\alpha,\beta} : \alpha, \beta \in \mathbb{N}^n, |\beta| = k\}$. For a function $f \in L^2(\mathbb{C}^n)$ we have the eigenfunction expansion

$$f = \sum_\alpha \sum_\beta (f, \Phi_{\alpha,\beta})\Phi_{\alpha,\beta}$$

which is called the special Hermite expansion. The above series converges in the L^2 norm.

The special Hermite series can be put in a compact form. We have to recall some basic facts about the Laguerre polynomials of type δ. They satisfy the generating function identity

$$\sum_{k=0}^{\infty} r^k L_k^{\delta}(t) e^{-\frac{1}{2}t} = (1-r)^{-\delta-1} e^{-\frac{1}{2}(\frac{1+r}{1-r})t}. \qquad (1.4.24)$$

From this and the formula for $\Phi_{\alpha,\alpha}$ follows the relation

$$\sum_{|\beta|=k} \Phi_{\beta,\beta}(z) = (2\pi)^{-\frac{n}{2}} L_k^{n-1}(\frac{1}{2}|z|^2) e^{-\frac{1}{4}|z|^2}. \qquad (1.4.25)$$

We define the Laguerre functions $\varphi_k(z)$ by

$$\varphi_k(z) = L_k^{n-1}(\frac{1}{2}|z|^2) e^{-\frac{1}{4}|z|^2}. \qquad (1.4.26)$$

We claim that the special Hermite expansions can be written in the compact form

$$f = (2\pi)^{-n} \sum_{k=0}^{\infty} f \times \varphi_k. \qquad (1.4.27)$$

To prove this claim let $Q_k f$ stand for the projection of f onto the eigenspace spanned by $\{\Phi_{\alpha,\beta}, \alpha, \beta \in \mathbb{N}^n, |\beta| = k\}$ so that

$$Q_k f = \sum_{|\beta|=k} \sum_{\alpha} (f, \Phi_{\alpha,\beta}) \Phi_{\alpha,\beta}.$$

If we can show that

$$f \times \Phi_{\beta,\beta} = \sum_{\alpha} (f, \Phi_{\alpha,\beta}) \Phi_{\alpha,\beta}$$

then our claim will be proved in view of the relation (1.4.25). To prove the latter formula we calculate the Weyl transform of $\Phi_{\alpha,\beta}$.

If $\varphi, \psi \in L^2(\mathbb{R}^n)$, then from the properties of the Fourier-Wigner transform, it follows that

$$(W(\bar{\Phi}_{\alpha,\beta})\varphi, \psi) = (2\pi)^{\frac{n}{2}} (V_{\varphi}(\psi), \Phi_{\alpha,\beta}) = (2\pi)^{\frac{n}{2}} (\varphi, \Phi_{\alpha})(\Phi_{\beta}, \psi).$$

Therefore,

$$(W(\bar{\Phi}_{\alpha,\beta} \times \bar{\Phi}_{\mu,\nu})\varphi, \psi) = (2\pi)^{\frac{n}{2}} (W(\bar{\Phi}_{\mu,\nu})\varphi, \Phi_{\alpha})(\Phi_{\beta}, \psi)$$

$$= (2\pi)^{\frac{n}{2}} (\varphi, \Phi_\mu)(\Phi_\nu, \Phi_\alpha)(\Phi_\beta, \psi)$$

and hence we get the interesting formula

$$\bar\Phi_{\alpha,\beta} \times \bar\Phi_{\mu,\nu} = (2\pi)^{\frac{n}{2}} \delta_{\nu,\alpha} \bar\Phi_{\mu,\beta};$$

as $\overline{f \times g} = \bar g \times \bar f$ we have

$$\Phi_{\alpha,\beta} \times \Phi_{\mu,\nu} = (2\pi)^{\frac{n}{2}} \delta_{\beta,\mu} \Phi_{\alpha,\nu}. \qquad (1.4.28)$$

The definition of Q_k together with the above formula proves our claim. From the above calculation it follows that

$$W(\Phi_{\alpha,\alpha})\varphi = (2\pi)^{\frac{n}{2}} (\varphi, \Phi_\alpha) \Phi_\alpha$$

and hence we have

$$W(\varphi_k)\varphi = (2\pi)^{\frac{n}{2}} \sum_{|\alpha|=k} (\varphi, \Phi_\alpha)\Phi_\alpha = (2\pi)^n P_k \varphi \qquad (1.4.29)$$

where P_k is the projection of $L^2(\mathbb{R}^n)$ onto the kth eigenspace spanned by $\{\Phi_\alpha : |\alpha| = k\}$. We thus have the interesting relation $W(\varphi_k) = (2\pi)^n P_k$ and hence

$$\varphi_k \times \varphi_j = (2\pi)^n \delta_{kj} \varphi_k. \qquad (1.4.30)$$

From this we also get the useful relation

$$W(Lf) = (2\pi)^n W(f)H$$

where L and H are the special Hermite and Hermite operators already introduced.

Using the above relations we will now show that the Weyl transform of a radial function reduces to the Laguerre transform. We will also write a formula for the group Fourier transform of a radial function on the Heisenberg group. Consider the Laguerre functions

$$\psi_k(r) = \left(\frac{2^{(1-n)}k!}{(k+n-1)!}\right)^{\frac{1}{2}} L_k^{n-1}\left(\frac{1}{2}r^2\right)e^{-\frac{1}{4}r^2}.$$

It is a fact that these functions form an orthonormal basis for the space $L^2(\mathbb{R}^+, r^{2n-1}dr)$ where $\mathbb{R}^+ = [0,\infty)$; see for example Szego [72]. If $f(z) = f(r), r = |z|$ is a radial function on $\mathbb{C}^n$ we can expand f as

$$f(r) = \sum_{k=0}^{\infty} \left(\int_0^\infty f(s)\psi_k(s)s^{2n-1}\,ds\right)\psi_k(r).$$

Recalling the definition of φ_k, this can be written as

$$f(z) = \sum_{k=0}^{\infty} R_k(f)\varphi_k(z)$$

where

$$R_k(f) = \frac{2^{(1-n)}k!}{(k+n-1)!} \int_0^{\infty} f(s)\varphi_k(s)s^{2n-1}\, ds. \qquad (1.4.31)$$

Using the orthogonality properties of the Laguerre functions under twisted convolution, we get

$$f \times \varphi_k = (2\pi)^n R_k(f)\varphi_k$$

and consequently the special Hermite expansion reduces to the Laguerre expansion. The above expansion leads to a simple formula for the Weyl transform of a radial function. Recalling that $W(\varphi_k) = (2\pi)^n P_k$, we have

$$W(f) = (2\pi)^n \sum_{k=0}^{\infty} R_k(f)P_k \qquad (1.4.32)$$

whenever f is a radial function. This result was first proved by Peetre [51].

The above formula can be pushed one step further to yield a formula for the group Fourier transform of a radial function on H^n. For functions on H^n, radial means radial in the z variable. To state the formula we introduce more notation. For every $\lambda \neq 0$, we define

$$\Phi_\alpha^\lambda(x) = |\lambda|^{\frac{n}{4}} \Phi_\alpha(|\lambda|^{\frac{1}{2}}x)$$

and let $P_k(\lambda)$ stand for the projections of $L^2(\mathbb{R}^n)$ onto the kth eigenspace spanned by $\{\Phi_\alpha^\lambda : |\alpha| = k\}$. We then have the following result.

Theorem 1.4.3 *Let f be a radial function on H^n. Then*

$$\hat{f}(\lambda) = \sum_{k=0}^{\infty} R_k(\lambda, f)P_k(\lambda)$$

where $R_k(\lambda, f) = R_k(f^\lambda)$.

Proof: If we let $\varphi_k^\lambda(z) = L_k^{n-1}(\frac{1}{2}|\lambda||z|^2)e^{-\frac{1}{4}|z|^2}$ then it is easy to see that

$$P_k(\lambda) = \int_{\mathbb{C}^n} \varphi_k^\lambda(z)\pi_\lambda(z)\,dz.$$

Expanding the function f^λ in terms of φ_k^λ and using the result for the Weyl transform, we can prove the theorem. We leave the details to the reader. ∎

Returning to the special Hermite expansions we know that the series converges in the L^2 sense. But we would like to know if the series converges pointwise for smooth and rapidly decreasing functions. More generally we are interested in the series when f is in the Schwartz class $S(\mathbb{C}^n)$. For such functions we prove the following result.

Theorem 1.4.4 *Finite linear combinations of special Hermite functions are dense in $S(\mathbb{C}^n)$.*

This theorem is proved in two steps. We first show that every Schwartz class function can be expanded in a mutiple Fourier series which will converge in the Schwartz topology. The terms of the Fourier series will have some homogeneity properties. We then show that for such homogeneous functions, the special Hermite series reduces to a multiple Laguerre series which will converge in the Schwartz topology. The two results combined will prove the above theorem.

First we set forth some definitions. Let $U(n)$ be the group of $n \times n$ unitary matrices. Then it is clear that a function f is radial if and only if $f(\sigma z) = f(z)$ for all $\sigma \in U(n)$. Now let $T(n)$ be the subgroup of $U(n)$ consisting of diagonal matrices. Each element of $T(n)$ can be written in the form $e^{i\theta} = (e^{i\theta_1}, \ldots, e^{i\theta_n})$ so that $T(n)$ is identified with the n-torus. The action of the torus on $\mathbb{C}^n$ is given by $e^{i\theta}z = (e^{i\theta_1}z_1, \ldots, e^{i\theta_n}z_n)$. Let m be an n-tuple of integers. We say that a function f is m-homogeneous if $f(e^{i\theta}z) = e^{im.\theta}f(z)$ for all θ. We say that the function is polyradial when it is 0-homogeneous. Equivalently, polyradial functions are precisely the functions which are invariant under the action of the torus and they are radial in each variable separately. We observe that when f is m-homogeneous and g is k-homogeneous, then $(f, g) = 0$ unless $k = m$. This follows by integrating in polar coordinates.

Given a function f on $\mathbb{C}^n$, we now define the m-radialisation by

$$R_m f(z) = (2\pi)^{-n} \int_Q f(e^{i\theta}z)e^{-im.\theta}\,d\theta \qquad (1.4.33)$$

where $Q = [0, 2\pi)^n$. Observe that $R_m f$ is m-homogeneous and we have the expansion

$$f(e^{i\theta}z) = \sum_m R_m f(z) e^{im\cdot\theta} \qquad (1.4.34)$$

which is the multiple Fourier series of the function $F_z(\theta) = f(e^{i\theta}z)$. Regarding the convergence of the above multiple Fourier series we now establish the following proposition.

Proposition 1.4.5 *If f is a Schwartz class function then the above series converges in the topology of $S(\mathbb{C}^n)$.*

Proof: For the sake of simplicity, let us assume $n = 1$. For $m \neq 0$ we integrate by parts to obtain

$$R_m f(z) = (im)^{-1}(2\pi)^{-1} \int_Q \frac{\partial}{\partial\theta} f(e^{i\theta}z) e^{-im\theta} \, d\theta.$$

Writing $z = x + iy$, we have

$$f(e^{i\theta}z) = f(x\cos\theta - y\sin\theta, y\cos\theta + x\sin\theta).$$

A simple calculation shows that

$$\frac{\partial}{\partial\theta} f(e^{i\theta}z) = N f(e^{i\theta}z)$$

where N is the rotation operator $(x\partial_y - y\partial_x)$. An iteration produces, for any integer k,

$$R_m f(z) = (im)^{-k}(2\pi)^{-1} \int_Q N^k f(e^{i\theta}z) e^{-im\theta} \, d\theta.$$

As f is in the Schwartz space, we get the estimates

$$|R_m f(z)| \leq C_k (1 + |m|)^{-k}$$

which shows that the series converges uniformly.

To show that the series actually converges in the Schwartz topology, we calculate the derivatives of $R_m f(z)$. We define the Cauchy-Riemann operators

$$\partial = \frac{1}{2}(\partial_x - i\partial_y), \quad \bar{\partial} = \frac{1}{2}(\partial_x + i\partial_y).$$

With this notation we have the relations

$$\partial_x R_m f = R_{m-1}(\partial f) + R_{m+1}(\bar{\partial} f), \quad \partial_y R_m f = i R_{m-1}(\partial f) - i R_{m+1}(\bar{\partial} f).$$

These relations give the formulas

$$\partial R_m f = R_{m-1}(\partial f), \quad \bar{\partial} R_m f = R_{m+1}(\bar{\partial} f).$$

In view of these formulas it is clear that $\partial^j R_m f$ and $\bar{\partial}^j R_m f$ are rapidly decreasing functions of $|m|$. We also have

$$(1 + |z|^2)^j R_m f(z) = (2\pi)^{-1} \int_0^{2\pi} (1 + |e^{i\theta} z|^2)^j f(e^{i\theta} z) e^{-i\theta} \, d\theta.$$

Hence it is clear that if q is a seminorm defining the topology of $S(\mathbb{C}^n)$ then

$$q(R_m f) \leq C(1 + |m|)^{-k}$$

for any k. Hence the series converges in the Schwartz topology. ∎

We now consider the special Hermite series of an m-homogeneous function. As $\Phi_{\mu,\nu}$ is $(\nu - \mu)-$ homogeneous , $(f, \Phi_{\mu,\nu}) = 0$ whenever $(\mu - \nu) \neq m$. Thus the projection $Q_k f$ reduces to the finite sum

$$Q_k f = \sum_{|\mu|=k} (f, \Phi_{\mu-m,\mu}) \Phi_{\mu-m,\mu},$$

and hence the special Hermite series reduces to a multiple Laguerre series.

Proposition 1.4.6 *If f is an m-homogeneous Schwartz class function, then its special Hermite series converges in the Schwartz topology.*

Proof: Recall that $\Phi_{\mu,\nu}$ are eigenfunctions of the operator L with eigenvalue $(2|\nu| + n)$. Therefore,

$$(f, \Phi_{\mu,\nu}) = (2|\nu| + n)^{-j}(f, L^j \Phi_{\mu,\nu})$$

$$= (2|\nu| + n)^{-j}(L^j f, \Phi_{\mu,\nu})$$

from which we get the estimate

$$|(f, \Phi_{\mu,\nu})| \leq C_j(f)(2|\nu| + n)^{-j}$$

for any j. Hence the series

$$\sum_{k=0}^{\infty} Q_k f = \sum_{k=0}^{\infty} \sum_{|\mu|=k} (f, \Phi_{\mu-m,\mu}) \Phi_{\mu-m,\mu},$$

converges uniformly.

Now let q be any seminorm defining the Schwartz topology. If we can show that $q(\Phi_{\mu-m,\mu})$ has only a polynomial growth in $|\mu|$, then it will follow that $q(Q_k f)$ has rapid decay as a function of k and consequently the series will converge in the Schwartz topology. To estimate $q(\Phi_{\mu-m,\mu})$, we need more properties of the Laguerre polynomials. We have the relation $\frac{d}{dt} L_k^{\delta}(t) = -L_{k-1}^{\delta+1}(t)$ and the estimate $|L_k^{\delta}(t)e^{-\frac{t}{2}}| \leq Ck^{\delta}$. Using these and the fact that $\Phi_{\mu-m,\mu}$ can be expressed in terms of Laguerre functions, we can prove the required estimates for $q(\Phi_{\mu-m,\mu})$. We leave the details.   ∎

Combining the two propositions above we get the theorem. Thus finite linear combinations of special Hermite functions are dense in $S(\mathbb{C}^n)$ and this fact will be used later in the study of mean periodic functions and spherical means.

1.5   Paley–Wiener theorems for the Fourier transform

The classical Paley-Wiener theorem for the Euclidean Fourier transform characterises compactly supported functions in terms of the behaviour of their Fourier transforms. In this section we are interested in establishing such results for the group Fourier transform and the Weyl transform. To motivate our intention, consider the Fourier transform of a function f on $\mathbb{R}^n$ translated by $\zeta \in \mathbb{R}^n$:

$$\mathcal{F}f(\xi + \zeta) = (2\pi)^{-\frac{n}{2}} \int_{\mathbb{R}^n} e^{-ix.(\xi+\zeta)} f(x)\, dx.$$

The Paley-Wiener theorem states that when f is compactly supported, we can take ζ to be complex and we have an estimate of the form

$$|\mathcal{F}f(\xi + \zeta)| \leq Ce^{B|Im(\zeta)|}$$

which measures the L^∞ norm of the translated Fourier transform as a function of $Im(\zeta)$. Therefore, a natural analogue of the Paley-Wiener theorem for the Heisenberg group will be to estimate bounded operators acting on $L^2(\mathbb{R}^n)$ translated by complex numbers ζ.

Now we have to make clear what we mean by a translated operator. To define this, let us recall the Weyl correspondence $f \to Weyl(f)$ by which an operator is associated to a function f on $\mathbb{C}^n$. This is just a variant of the Weyl transform, namely,

$$Weyl(f) = W(\mathcal{F}^{-1}f);$$

that is, $Weyl(f)$ is the Weyl transform of the inverse Fourier transform of f. Let $\tau(u,v)$ be the translation operator $\tau(u,v)f(x,y) = f(x-u, y-v)$. Then we have the relation

$$Weyl(\tau(u,v)f) = W(e^{i(x.u+y.v)}\mathcal{F}^{-1}f).$$

In view of the formula

$$\pi(x+iy)\pi(u+iv) = e^{\frac{i}{2}(u.y-v.x)}\pi((x+u)+i(y+v))$$

we get the relation

$$Weyl(\tau(u,v)f) = \pi(-v+iu)Weyl(f)\pi(v-iu).$$

Therefore, operators of the form $\pi(u+iv)S\pi(-u-iv)$ can be considered as translations of the operator S by $(u,v) \in \mathbb{R}^{2n}$. With this definition of translation, we want to take $S = \hat{f}(\lambda)$, translate it by $\xi \in \mathbb{R}^{2n}$, and see if this translation can be defined for $\zeta \in \mathbb{C}^{2n}$. If so we would like to know how the operator norm of the translated operator behaves as a function of ζ. Then we proceed as follows.

We first consider the Weyl transform. If $f \in L^2(\mathbb{C}^n)$ then we know that $W(f)$ is a Hilbert-Schmidt operator. We now embed $W(f)$ in a family of Hilbert-Schmidt operators in the following way. For $\xi = (\xi', \xi'') \in \mathbb{R}^{2n}$ let us write $U(\xi) = \pi(\xi' + i\xi'')$. To every $f \in L^2(\mathbb{R}^n)$ we now define the Fourier-Weyl transform by

$$\tilde{f}(\xi) = U(\xi)W(f)U(-\xi). \tag{1.5.35}$$

Since $U(\xi)$ are unitary operators, we note that $\tilde{f}(\xi) \in S_2$ for every $\xi \in \mathbb{R}^{2n}$. The image of $L^2(\mathbb{C}^n)$ under the Fourier-Weyl transform thus consists of functions $F(\xi)$ taking values in S_2 which verify the relation

$$F(0) = U(-\xi)F(\xi)U(\xi).$$

We now let E_0 stand for the subspace of this image whose elements are restrictions to $\mathbb{R}^{2n}$ of entire functions of exponential type taking values in S_2. In other words $F \in E_0$ if and only if (i) $F(\zeta)$ is an entire function of ζ taking values in S_2 which satisfies the estimate $\|F(\zeta)\|_{HS} \leq Ce^{B|Im(\zeta)|}$ for some constant $B > 0$ and (ii)$F(0) = U(-\xi)F(\xi)U(\xi)$ for all $\xi \in \mathbb{R}^{2n}$.

The space E_0 can be equipped with a topology as follows. Let $E'(\mathbb{R}^{2n}, S_2)$ be the space of compactly supported distributions on $\mathbb{R}^{2n}$ taking values in the Hilbert space S_2. That is, any $T \in E'(\mathbb{R}^{2n}, S_2)$ is a continuous linear operator from $C^\infty(\mathbb{R}^{2n})$ into S_2 where $C^\infty(\mathbb{R}^{2n})$ is equipped with the topology of uniform convergence of all derivatives on compact sets. To every $T \in E'(\mathbb{R}^{2n}, S_2)$, we associate its Fourier transform $\hat{T}(\xi) = (T, e_\xi)$ where $e_\xi(x) = e^{ix\cdot\xi}$. The Paley-Wiener theorem for compactly supported Hilbert space valued distributions says that $T \in E'(\mathbb{R}^{2n}, S_2)$ if and only if $\hat{T}(\xi)$ extends to an entire function of exponential type satisfying

$$\|\hat{T}(\zeta)\|_{HS} \leq C(1 + |\zeta|)^N e^{B|Im(\zeta)|}$$

for some constants N and B . Let E stand for the image of $E'(\mathbb{R}^{2n}, S_2)$ under the Fourier transform.

The space E is equipped with the strong topology which makes the Fourier transform a topological isomorphism between $E'(\mathbb{R}^{2n}, S_2)$ and E. We refer to Treves [75] for a discussion of topologies on various spaces of test functions and distributions. In this topology a sequence F_j converges to F iff the following two things happen: (i) $F_j(\zeta) \to F(\zeta)$ uniformly on compact sets (ii) $F_j(\zeta)$ and $F(\zeta)$ verify the estimates with N and B independent of j. The space E_0 inherits a topology from E.

Proposition 1.5.1 E_0 *is a closed subspace of* E.

Proof: Suppose $F_j(\zeta) \in E_0$ and $F_j(\zeta) \to F(\zeta)$ in E. From the definition it is clear that $F_j(0) \to F(0)$ in S_2 and hence $F(0) = W(f)$ for some $f \in L^2(\mathbb{C}^n)$. We need to show that $W(f) = U(-\xi)F(\xi)U(\xi)$ for all $\xi \in \mathbb{R}^{2n}$. To see this we observe that

$$U(-\xi)F_j(\xi)U(\xi) - F(0) = U(-\xi)(F_j(\xi) - U(\xi)F(0)U(-\xi))U(\xi).$$

This means that

$$F_j(\xi) - U(\xi)F(0)U(-\xi) = U(\xi)(F_j(0) - F(0))U(-\xi)$$

and consequently $F_j(\xi) - U(\xi)F(0)U(-\xi)$ converges to 0 in S_2. But then we should have $F(\xi) = U(\xi)F(0)U(-\xi)$ which proves the claim. ∎

We now state a Paley-Wiener theorem for the Weyl transform. Let $E'(\mathbb{C}^n)$ be the space of compactly supported distributions on $\mathbb{C}^n$ and let $L_0^2(\mathbb{C}^n) = L^2 \cap E'(\mathbb{C}^n)$ be the set of all compactly supported $f \in L^2(\mathbb{C}^n)$.

Theorem 1.5.2 *The Fourier-Weyl transform sets up an isomorphism between $L_0^2(\mathbb{C}^n)$ and E_0; that is $f \in L_0^2(\mathbb{C}^n)$ if and only if its Fourier-Weyl transform $\tilde{f} \in E_0$.*

Proof: One easily verifies that

$$U(\xi)\pi(z)U(-\xi) = e^{i(x.\xi'' - y.\xi')}\pi(z)$$

and consequently

$$\tilde{f}(\xi) = \int_{\mathbb{C}^n} e^{i(x.\xi'' - y.\xi')} f(z)\pi(z)\, dz.$$

If now f is supported in $|z| \leq B$ then $\tilde{f}(\xi)$ can be extendeded to an entire function $\tilde{f}(\zeta)$ of $\zeta \in \mathbb{C}^{2n}$, and by the Plancherel theorem for the Weyl transform, we also have the estimate

$$\|\tilde{f}(\zeta)\|_{HS} \leq Ce^{B|Im(\zeta)|}.$$

This proves the direct part. To prove the converse, we proceed as follows.

Let $F(\zeta) \in E_0$ and let $f \in L^2(\mathbb{C}^n)$ be such that $F(0) = W(f)$. We need to show that f is compactly supported. Consider for $\xi' \in \mathbb{R}^n$ the function

$$\tilde{f}_1(\xi') = F(\xi', 0) = \pi(\xi')W(f)\pi(-\xi').$$

Then a calculation shows that

$$\tilde{f}_1(\xi') = \int_{\mathbb{C}^n} e^{-iy.\xi'} f(z)\pi(z)\, dz.$$

Suppose $F(\zeta)$ satisfies $\|F(\zeta)\|_{HS} \leq Ce^{B|Im(\zeta)|}$. Let $\varphi, \psi \in L^2(\mathbb{R}^n)$ and consider the function

$$(\tilde{f}_1(\xi')\varphi, \psi) = \int_{\mathbb{C}^n} e^{-iy.\xi'} f(z)(\pi(z)\varphi, \psi)\, dz.$$

This is an entire function of exponential type. If we let

$$g(y) = \int_{\mathbb{C}^n} f(z)(\pi(z)\varphi, \psi)\, dx,$$

then $(\tilde{f}_1(\xi')\varphi, \psi)$ is the Fourier transform of g, and hence by the classical Paley-Wiener theorem, it follows that g is supported in $|y| \leq B$. From this we want to conclude that f itself vanishes for $|y| \geq B$.

To prove this we take $\varphi = \Phi_0$ and $\psi = \Phi_m$ where m is a multi-index. Then we know that

$$(\pi(z)\varphi, \psi) = c_m z^m e^{-\frac{1}{4}|z|^2}$$

where c_m is a constant. Therefore, when $|y| \geq B$ we have

$$\int_{\mathbb{R}^n} f(z)(x+iy)^m e^{-\frac{1}{4}|z|^2} \, dx = 0.$$

Since this is true for all m, we see that $f(x+iy)$ is orthogonal to all Hermite functions and hence $f(x+iy) = 0$ for all x whenever $|y| \geq B$. Similarly by considering $F(0, \xi'')$ we can show that $f(x+iy)$ is supported in $|x| \leq B$. This completes the proof of the theorem. ∎

We can formulate a similar theorem for the group Fourier transform which characterises functions that are compactly supported in the z variable. Since we are interested in a Paley-Wiener theorem which respects both variables, we will not go into the formulation of this result. Instead, we will rephrase the above theorem in a slightly different form which gives us enough motivation for what we are going to do in the case of the group Fourier transform.

Again we return to the Fourier transform on $\mathbb{R}^n$. Taking derivatives of $\hat{f}$, we have the relation

$$\partial_\xi^\alpha \hat{f}(\xi) = (2\pi)^{-\frac{n}{2}} \int_{\mathbb{R}^n} e^{-ix\cdot\xi}(-ix)^\alpha f(x) \, dx \qquad (1.5.36)$$

for any multiindex α. If f is compactly supported in $|x| \leq B$, then it is clear that

$$|\partial_\xi^\alpha \hat{f}(\xi)| \leq CB^{|\alpha|}$$

for all α. The converse is also true. If the above estimates are true, then the function

$$F(\zeta) = \sum_\alpha \frac{\partial_\xi^\alpha \hat{f}(0)}{\alpha!} \zeta^\alpha$$

defines an entire function and as

$$F(\xi + i\eta) = \sum_\alpha \frac{\partial_\xi^\alpha \hat{f}(\xi)}{\alpha!}(i\eta)^\alpha$$

it also satisfies the estimate $|F(\zeta)| \leq Ce^{B|Im(\zeta)|}$. Hence by the Paley-Wiener theorem for the Fourier transform, f will be supported in $|x| \leq B$. We look for a similar result in the case of the Weyl transform.

Consider the multiplication operators M_j and $\bar{M}_j, j = 1, 2, \ldots, n$ defined by

$$M_j f(z) = z_j f(z), \quad \bar{M}_j f(z) = \bar{z}_j f(z). \qquad (1.5.37)$$

We calculate the Weyl transform of $M_j f$ and $\bar{M}_j f$ in terms of $W(f)$. We introduce the following derivations. Let

$$A_j = \frac{\partial}{\partial \bar{\xi}_j} + \xi_j, \quad A_j^* = -\frac{\partial}{\partial \xi_j} + \xi_j \qquad (1.5.38)$$

be the annihilation and creation operators of quantum mechanics. For a bounded operator T on $L^2(\mathbb{R}^n)$, we define the derivations

$$\delta_j T = [A_j, T] = A_j T - T A_j,$$

$$\bar{\delta}_j T = -[A_j^*, T] = T A_j^* - A_j^* T.$$

It is easy to see that these are derivations in the sense that $\delta_j(TS) = (\delta_j T)S + T(\delta_j S)$. A simple calculation shows that $W(M_j f) = \bar{\delta}_j W(f)$ and $W(\bar{M}_j f) = \delta_j W(f)$. By iteration we get

$$\bar{\delta}_j^\alpha W(f) = W(z^\alpha f), \quad \delta_j^\alpha W(f) = W(\bar{z}^\alpha f) \qquad (1.5.39)$$

where $\bar{\delta}_j^\alpha$ and δ_j^α are defined in the obvious way. We now have the following version of the Paley-Wiener theorem.

Theorem 1.5.3 *Assume that $1 \leq p \leq 2$ and $f \in L^p(\mathbb{C}^n)$. Then f is supported in $|z| \leq B$ if and only if $W(f)$ satisfies $\|\delta_j^\alpha \bar{\delta}_j^\beta W(f)\| \leq CB^{|\alpha|+|\beta|}$ for any α and β.*

Proof: Let $\varphi, \psi \in L^2(\mathbb{R}^n)$. Then we know that their Fourier-Wigner transform $V_\varphi(\psi) \in L^q(\mathbb{C}^n)$ for all $q \geq 2$. Therefore,

$$(W(f)\varphi, \psi) = (2\pi)^{\frac{n}{2}} \int_{\mathbb{C}^n} f(z) V_\varphi(\psi, z) \, dz$$

shows that $W(f)$ is a bounded operator on $L^2(\mathbb{R}^n)$ whenever $f \in L^p(\mathbb{R}^n)$. If f is supported in $|z| \leq B$, then using (1.5.39), we get

$$\|\delta_j^\alpha \bar{\delta}_j^\beta W(f)\| \leq C\|f\|_p B^{|\alpha|+|\beta|}$$

and this proves the direct part of the theorem.

To prove the converse, we look at the Fourier-Weyl transform

$$\tilde{f}(\xi) = \int_{\mathbb{C}^n} e^{i(x \cdot \xi'' - y \cdot \xi')} f(z) \pi(z) \, dz.$$

Differentiating with respect to ξ_j', we get

$$\frac{\partial}{\partial \xi_j'} \tilde{f}(\xi) = \int_{\mathbb{C}^n} e^{i(x \cdot \xi'' - y \cdot \xi')} \frac{1}{2} (\bar{z}_j - z_j) f(z) \pi(z) \, dz.$$

In view of the relations between the derivations and the multiplications by coordinate functions, we get

$$\frac{\partial}{\partial \xi_j'} \tilde{f}(\xi) = \frac{1}{2} U(\xi)(\delta_j - \bar{\delta}_j) W(f) U(-\xi).$$

Similarly, we can show that

$$\frac{\partial}{\partial \xi_j''} \tilde{f}(\xi) = \frac{1}{2} U(\xi)(\delta_j + \bar{\delta}_j) W(f) U(-\xi).$$

Iteration produces the result

$$\partial_{\xi''}^\beta \partial_{\xi'}^\alpha \tilde{f}(\xi) = 2^{-(|\alpha|+|\beta|)} U(\xi)(\delta + \bar{\delta})^\beta (\delta - \bar{\delta})^\alpha W(f) U(-\xi).$$

Under the hypothesis on the derivatives of $W(f)$, we get the estimates

$$\|\partial_{\xi''}^\beta \partial_{\xi'}^\alpha \tilde{f}(\xi)\| \le C B^{|\alpha|+|\beta|}.$$

Consequently, for each $\zeta = (\zeta', \zeta'') \in \mathbb{C}^{2n}$, the series

$$F(\zeta) = \sum_{\alpha, \beta} \frac{\partial_{\xi''}^\beta \partial_{\xi'}^\alpha \tilde{f}(0)}{\alpha! \beta!} \zeta'^\alpha \zeta''^\beta$$

converges and defines an operator valued entire function. Moreover, since

$$F(\xi + i\eta) = \sum_{\alpha, \beta} \frac{\partial_{\xi''}^\beta \partial_{\xi'}^\alpha \tilde{f}(\xi', \xi'')}{\alpha! \beta!} (i\eta')^\alpha (i\eta'')^\beta,$$

we have the estimate

$$\|F(\zeta)\| \le C e^{B|Im(\zeta)|}.$$

Therefore, by the previous theorem we conclude that f is supported in $|z| \leq B$. This completes the proof of the theorem. ∎

We will now formulate and prove a Paley-Wiener theorem for the group Fourier transform on the Heisenberg group. First we have to recall a few definitions from the theory of Lie algebras. Each representation π of the Heisenberg group defines a skew adjoint representation, also denoted by π, of the Heisenberg Lie algebra h_n by the formula

$$\pi(X)\varphi = \frac{d}{dt}|_{t=0}\pi(\exp(tX))\varphi$$

where $\exp : h_n \to H^n$ is the exponential map and φ is a C^∞ vector for π. The skew adjointness of π means that $\pi(X)^* = -\pi(X)$.

Recall that the Heisenberg Lie algebra is generated by the left invariant vector fields $X_1, \ldots, X_n, Y_1, \ldots, Y_n$ and T. Let us write $X_{j+n} = Y_j, j = 1, 2, \ldots, n$. It then follows that if for each $\xi \in \mathbb{R}^{2n}$ we define

$$U_\lambda(\xi) = \exp(-\sum_{j=1}^{n} \xi_j \pi_\lambda(X_j)) \qquad (1.5.40)$$

then $U_\lambda(\xi)$ becomes a unitary operator. We use this operator valued function in formulating the Paley-Wiener theorem.

Let $\tilde{X}_j, j = 1, 2, \ldots, 2n$ be the right invariant vector fields agreeing with X_j at the origin. An easy calculation shows that

$$\tilde{X}_j = \frac{\partial}{\partial x_j} + \frac{1}{2}y_j\frac{\partial}{\partial t}, \quad \tilde{X}_{n+j} = \frac{\partial}{\partial y_j} - \frac{1}{2}x_j\frac{\partial}{\partial t} \qquad (1.5.41)$$

for $j = 1, 2, \ldots, 2n$. It then follows that

$$\tilde{X}_j - X_j = y_j\frac{\partial}{\partial t}, \quad \tilde{X}_{n+j} - X_{n+j} = -x_j\frac{\partial}{\partial t}$$

for $j = 1, 2, \ldots, n$. If π_λ are the Schrödinger representations, then a direct calculation shows that

$$\pi_\lambda(-Xf) = \pi_\lambda(f)\pi_\lambda(X), \quad \pi_\lambda(\tilde{X}f) = -\pi_\lambda(X)\pi_\lambda(f)$$

for any left invariant vector field X. Here $\pi_\lambda(f)$ stands for $\hat{f}(\lambda)$ in our notation. The last two equations imply that

$$\pi_\lambda((\tilde{X} - X)f) = [\pi_\lambda(f), \pi_\lambda(X)].$$

Thus, as in the case of the Weyl transform, we are led to the following derivations.

$$\delta_j(\lambda)T = [T, \pi_\lambda(X)], \quad j = 1, 2, \ldots, 2n.$$

We say that an operator T is of class C^k if $\delta(\lambda)^\alpha T$ is bounded for all α with $|\alpha| \le k$. Let us write

$$P_j(z,t) = y_j t, \quad P_{j+n}(z,t) = -x_j t, \quad j = 1, 2, \ldots, n$$

so that

$$(\tilde{X}_j - X_j)\tilde{f}(z,t) = -i(P_j f)\tilde{\ }(z,t).$$

Here $\tilde{f}$ stands for the Fourier transform of f in the t variable. Applying π_λ to both sides of the above equation, we get

$$\delta_j(\lambda)\pi_\lambda(\tilde{f}) = -i\pi_\lambda((P_j f)\tilde{\ })$$

for $j = 1, 2, \ldots, 2n$. We now define the modified Fourier transform in the following way. For each $\xi \in \mathbb{R}^{2n}, \hat{f}(\xi)$ takes values in the Hilbert space $K = L^2(\mathbb{R}^*, S_2; d\mu)$ and is given by

$$\hat{f}(\xi)(\lambda) = U_\lambda(\xi)\pi_\lambda(\tilde{f})U_\lambda(-\xi).$$

Recalling the definition of $U_\lambda(\xi)$, taking derivatives with respect to ξ_j and using (1.5.40) we obtain the interesting formula

$$\frac{\partial}{\partial \xi_j}\hat{f}(\xi) = -i(P_j f)\hat{\ }(\xi) \qquad\qquad (1.5.42)$$

which is the analogue of (1.5.36) for the modified Fourier transform.

We can now state and prove a Paley-Wiener theorem for our modified Fourier transform. Let $H_B(\mathbb{C}^{2n})$ stand for the space of entire functions $F(\zeta)$ taking values in K which agree with $\hat{f}(\xi)$ on $\mathbb{R}^{2n}$ for some $f \in S(H^n)$ and verify the estimate

$$\|F(\zeta)\|_K \le Ce^{B|Im(\zeta)|}.$$

Let G_B stand for the set defined by

$$\{(z,t) : |P_j(z,t)| \le B, \ j = 1, 2, \ldots, 2n\}$$

and let $C^\infty(G_B)$ be the space of smooth functions supported in G_B.

Theorem 1.5.4 *The modified Fourier transform sets up an isomorphism between the spaces $C^\infty(G_B) \cap S(H^n)$ and $H_B(\mathbb{C}^{2n})$.*

Proof: The direct part of this theorem is easy to prove. If a Schwartz class function f is supported in G_B, then an iteration of (1.5.42) gives us

$$\partial_\xi^\alpha \hat{f}(\xi) = (-i)^{|\alpha|}(P^\alpha f)\hat{}(\xi) \qquad (1.5.43)$$

where $P^\alpha(z,t) = P_1(z,t)^{\alpha_1} P_2(z,t)^{\alpha_2}....P_{2n}(z,t)^{\alpha_{2n}}$. It then follows that

$$\|\partial_\xi^\alpha \hat{f}(\xi)\|_K^2 = \int_{-\infty}^\infty \|\pi_\lambda((P^\alpha f)\tilde{})\|_{HS}^2 \, d\mu(\lambda) \qquad (1.5.44)$$

which by Plancherel theorem gives

$$\|\partial_\xi^\alpha \hat{f}(\xi)\|_K^2 = \int_{H^n} |P^\alpha(z,t) f(z,t)|^2 \, dzdt.$$

As f is supported in G_B, we get

$$\|\partial_\xi^\alpha \hat{f}(\xi)\|_K^2 \le B^{2|\alpha|} \int_{H^n} |f(z,t)|^2 \, dzdt \le CB^{2|\alpha|}.$$

From these estimates it follows that the series

$$F(\zeta) = \sum \frac{1}{\alpha!} \partial_\xi^\alpha \hat{f}(0)\zeta^\alpha$$

converges in the norm of K and defines an entire function. Moreover, the expansion

$$F(\xi + i\eta) = \sum \frac{1}{\alpha!} \partial_\xi^\alpha \hat{f}(\xi)(i\eta)^\alpha$$

shows that we also have the estimate $\|F(\zeta)\|_K \le Ce^{B|Im(\zeta)|}$. This shows that $F(\zeta) \in H_B(\mathbb{C}^{2n})$.

We now turn to the converse. Let $F \in H_B(\mathbb{C}^{2n})$ and let f be a Schwartz class function such that $\hat{f}(\xi) = F(\xi), \xi \in \mathbb{R}^{2n}$. We need to show that f is supported in G_B. This will follow immediately if we can show that

$$\int_{H^n} |P_j(z,t)|^{2k} |f(z,t)|^2 \, dzdt \le B^{2k} \int_{H^n} |f(z,t)|^2 \, dzdt \qquad (1.5.45)$$

for all k. Again in view of (1.5.45) it is enough to show that

$$\|(\frac{\partial}{\partial \xi_j})^k \hat{f}(0)\|_K^2 \le B^{2k} \|f\|_2^2.$$

In order to establish this we proceed as follows.

Let $\theta \in C_0^\infty(\mathbb{R}^{2n})$ be a real valued function supported in $|x| \leq 1$ and $\|\theta\|_2 = 1$. For each $\epsilon > 0$ let $\theta_\epsilon(x) = \epsilon^{-n}\theta(\frac{x}{\epsilon})$ so that $\hat{\theta}_\epsilon(\xi) = \epsilon^n\hat{\theta}(\epsilon\xi)$. By the Paley-Wiener theorem, we know that $\hat{\theta}_\epsilon$ extends to an entire function which satisfies the estimate

$$|\hat{\theta}_\epsilon(\zeta)| \leq C_\epsilon e^{\epsilon|Im(\zeta)|}.$$

As $\theta \in C_0^\infty(\mathbb{R}^{2n})$ we also know that $\hat{\theta}_\epsilon \in L^2(\mathbb{R}^{2n})$. We now consider the function $M_\epsilon(\zeta) = \hat{\theta}_\epsilon(\zeta)F(\zeta)$. This is an entire function taking values in K which satisfies

$$\|M_\epsilon(\zeta)\|_K \leq C_\epsilon e^{(B+\epsilon)|Im(\zeta)|}.$$

Moreover, the calculation

$$\int_{\mathbb{R}^{2n}} \|M_\epsilon(\xi)\|_K^2 \, d\xi = \|f\|_2^2 \int_{\mathbb{R}^{2n}} |\theta_\epsilon(\xi)|^2 \, d\xi = \|f\|_2^2$$

shows that $M_\epsilon \in L^2(\mathbb{R}^{2n}, K)$. Now we can appeal to the Paley-Wiener theorem to conclude that there is a function T_ϵ in the space $L^2(\mathbb{R}^{2n}, K)$ supported in $|x| \leq (B + \epsilon)$ such that

$$M_\epsilon(\xi) = \int_{\mathbb{R}^{2n}} e^{-ix.\xi} T_\epsilon(x) \, dx.$$

Differentiating this k times with respect to ξ_j, we get

$$\left(\frac{\partial}{\partial\xi_j}\right)^k M_\epsilon(\xi) = \int_{\mathbb{R}^{2n}} e^{-ix.\xi}(-ix_j)^k T_\epsilon(x) \, dx.$$

By the Euclidean Plancherel theorem we have

$$\int_{\mathbb{R}^{2n}} \|(\frac{\partial}{\partial\xi_j})^k M_\epsilon(\xi)\|_K^2 = \int_{\mathbb{R}^{2n}} \|x_j^k T_\epsilon(x)\|_K^2 \, dx.$$

As $T_\epsilon(x)$ is supported in $|x| \leq (B + \epsilon)$, we have

$$\int_{\mathbb{R}^{2n}} \|x_j^k T_\epsilon(x)\|_K^2 \, dx \leq (B + \epsilon)^{2k} \int_{\mathbb{R}^{2n}} \|T_\epsilon(x)\|_K^2 \, dx$$

and by Plancherel theorem again

$$\int_{\mathbb{R}^{2n}} \|T_\epsilon(x)\|_K^2 \, dx = \int_{\mathbb{R}^{2n}} \|M_\epsilon(\xi)\|_K^2 \, d\xi.$$

Thus we get the estimate

$$\int_{\mathbb{R}^{2n}} \|(\frac{\partial}{\partial \xi_j})^k M_\epsilon(\xi)\|_K^2 \leq (B + \epsilon)^{2k} \|f\|_2^2.$$

Since $M_\epsilon(\xi)$ is the product of $F(\xi)$ and $\hat{\theta}_\epsilon$ by the Leibniz formula, we have the relation

$$\partial_j^k M_\epsilon(\xi) = \sum_{l=0}^{k} A_k^l \partial_j^l \hat{\theta}_\epsilon(\xi) \partial_j^{k-l} F(\xi)$$

where $\partial_j = \frac{\partial}{\partial \xi_j}$. From the above we calculate

$$\|\partial_j^k M_\epsilon(\xi)\|_{HS}^2$$

$$= \sum_{l=0}^{k} \sum_{i=0}^{k} A_k^l A_k^i \partial_j^l \hat{\theta}_\epsilon(\xi) \partial_j^i \hat{\theta}_\epsilon(\xi) tr((\partial_j^{k-l} \hat{f}(\xi))(\partial_j^{k-i} \hat{f}(\xi))^*).$$

But

$$tr((\partial_j^{k-l} \hat{f}(\xi))(\partial_j^{k-i} \hat{f}(\xi))^* = tr((\partial_j^{k-l} \hat{f}(0))(\partial_j^{k-i} \hat{f}(0))^*)$$

and we have the inequality

$$\sum_{l=0}^{k} \sum_{i=0}^{k} A_k^l A_k^i \{ \int_{\mathbb{R}^{2n}} \partial_j^l \hat{\theta}_\epsilon(\xi) \partial_j^i \hat{\theta}_\epsilon(\xi) \, d\xi \}$$

$$\times \{ \int_{\mathbb{R}^*} tr((\partial_j^{k-l} \hat{f}(0))(\partial_j^{k-i} \hat{f}(0))^*) \, d\mu \} \leq (B + \epsilon)^{2k} \|f\|_2^2.$$

As $\hat{\theta}_\epsilon(\xi) = \epsilon^n \hat{\theta}(\epsilon \xi)$ it follows that the integral

$$\int_{\mathbb{R}^{2n}} \partial_j^l \hat{\theta}_\epsilon(\xi) \partial_j^i \hat{\theta}_\epsilon(\xi) \, d\xi = \epsilon^{l+i} \int_{\mathbb{R}^{2n}} \partial_j^l \hat{\theta}(\xi) \partial_j^i \hat{\theta}(\xi) \, d\xi$$

converges to 0 as $\epsilon \to 0$ unless $l = 0, i = 0$. Therefore if we let $\epsilon \to 0$ in the above, only one term survives, and we have

$$\int_{\mathbb{R}^*} tr((\partial_j^k \hat{f}(0))(\partial_j^k \hat{f}(0))^*) \, d\mu \leq B^{2k} \|f\|_2^2.$$

This completes the proof of the theorem. ∎

1.6   An uncertainty principle on the Heisenberg group

The uncertainty principle for the Fourier transform on $\mathbb{R}^n$ states that if a function f is concentrated, then its Fourier transform cannot be concentrated unless f is identically zero. There are several ways of measuring the concentration and consequently there are several formulations of the uncertainty principle; see the survey article of Folland-Sitaram [28]. The simplest example is the following consequence of the Paley-Wiener theorem: if a function f is compactly supported then its Fourier transform cannot have compact support unless $f = 0$. Some other measures of concentration are: the quadratic mean variation of the function from a point, the Lebesgue measure of the support of the function and the decay of the function at infinity, to mention only a few. A classical theorem of Hardy on Fourier transform pairs says that if a function f and its Fourier transform satisfy the estimates

$$|f(t)| \le Ce^{-at^2}, \ |\hat{f}(s)| \le Ce^{-bs^2},$$

for some positive constants a and b then $f = 0$ whenever $ab > \frac{1}{4}$. When $ab < \frac{1}{4}$, there are infinitely many linearly independent functions satisfying the above estimates and when $ab = \frac{1}{4}, f(t) = Ce^{-at^2}$. A proof of this theorem can be found in Dym-McKean [19].

Uncertainty principles seem to be a general feature of harmonic analysis on Lie groups. On the Heisenberg group we have several forms of the uncertainty principle. As the group Fourier transform $\hat{f}(\lambda)$ is given by integrating f^λ against π_λ, an immediate consequence of the Paley-Wiener theorem for the Fourier transform in the t variable is the following: if f is compactly supported, then the support of $\hat{f}(\lambda)$ (as an operator valued function) in $\mathbb{R}^*$ cannot be compact unless $f = 0$. Since the condition $\hat{f}(\lambda) = 0$ imposes strong restrictions on the function ($\hat{f}(\lambda) = 0$ forces $f^\lambda = 0$), we look for versions of uncertainty principles which impose comparatively weaker conditions on the Fourier transform.

In this section we are mainly interesteted in proving analogues of Hardy's theorem for the Fourier and Weyl transforms. We start with a Hardy's theorem for the Fourier transform on $\mathbb{R}^n$ which we need in order to establish Hardy's theorem for the Weyl transform.

Theorem 1.6.1 *Let f be a measurable function on $\mathbb{R}^n$ and let a and b be two positive constants. Assume that*

$$|f(x)| \le Ce^{-a|x|^2}, \ |\hat{f}(\xi)| \le Ce^{-b|\xi|^2},$$

for all x and ξ in $\mathbb{R}^n$. Then $f = 0$ whenever $ab > \frac{1}{4}$. There are infinitely many linearly independent functions satisfying the above estimates when $ab < \frac{1}{4}$ and when $ab = \frac{1}{4}$ the function has to be a constant multiple of the Gaussian $e^{-a|x|^2}$.

Proof: The one dimensional case of Hardy's theorem is proved using the Phragmen-Lindelof argument [19] . We assume the case $n = 1$ and reduce the higher dimensional case to the one-dimensional one by using the Radon transform. Recall that the Radon transform Rg of a function g on $\mathbb{R}^n$ is a function on $S^{n-1} \times \mathbb{R}$ and is given by

$$Rg(\omega, s) = \int_{x.\omega=s} g(x) \, dm(x)$$

where $dm(x)$ is the Lebesgue measure on the hyperplane $x.\omega = s$. For each fixed ω, the above makes sense for almost every $s \in \mathbb{R}$. There is an inversion theorem for the Radon transform: if $Rg(\omega, s) = 0$ for almost every $(\omega, s) \in S^{n-1} \times \mathbb{R}$ then $g(x) = 0$ for almost every x. For basic properties of the Radon transform, we refer to Folland [27].

The Radon transform and the Fourier transform are related. An easy calculation shows that

$$\mathcal{F}f(s\omega) = (R\tilde{f})(\omega, s)$$

where $(R\tilde{f})$ is the Fourier transform of Rf in the t variable. In view of this relation the conditions on f and $\hat{f}$ translate into the conditions

$$|Rf(\omega, t)| \leq Ce^{-at^2}, \ |(R\tilde{f})(\omega, s)| \leq Ce^{-bs^2}.$$

Applying the one-dimensional Hardy theorem, we conclude that if $ab > \frac{1}{4}$ then $Rf(\omega, .) = 0$ for almost all ω. By the inversion theorem for the Radon transform, this implies $f = 0$. When $ab = \frac{1}{4}$, $(R\tilde{f})(\omega, s) = \hat{f}(\omega s) = g(\omega)e^{-bs^2}$ where g is a function on S^{n-1}. As f is integrable, $\hat{f}$ is continuous and so by taking limit as $s \to 0$, we get $g(\omega) = \hat{f}(0)$. Hence $\hat{f}(\xi) = \hat{f}(0)e^{-b|\xi|^2}$ so that $f(x) = Ce^{-a|x|^2}$. Finally when $ab < \frac{1}{4}$, the n dimensional Hermite functions Φ_α satisfy the estimates of the theorem. ∎

Using Hardy's theorem for the Fourier transform, we now prove the following analogue for the group Fourier transform on the Heisenberg group.

Theorem 1.6.2 *Let f be a measurable function which satisfies the estimates $|f(z,t)| \leq g(z)e^{-at^2}$ where $g \in L^1 \cap L^2(\mathbb{C}^n)$ and $\|\hat{f}(\lambda)\|_{HS} \leq Ce^{-b\lambda^2}$ with some positive constants a and b. Then $f = 0$ whenever $ab > \frac{1}{4}$. There are infinitely many linearly independent functions satisfying the above estimates when $ab < \frac{1}{4}$.*

Proof: We prove the theorem by reducing it to the Euclidean case. Define the function $f^*(z,t) = \bar{f}(z,-t)$ and let $f *_3 f^*$ be the convolution of f and f^* in the t variable. Then a simple calculation shows that

$$\int_{H^n} f *_3 f^*(z,t)e^{i\lambda t}\, dz dt$$

$$= \int_{\mathbb{C}^n} (f *_3 f^*)^\lambda(z)\, dz = \int_{\mathbb{C}^n} |f^\lambda(z)|^2\, dz.$$

But we already know that

$$\int_{\mathbb{C}^n} |f^\lambda(z)|^2\, dz = (2\pi)^n |\lambda|^n \|\hat{f}(\lambda)\|_{HS}^2.$$

So if we let

$$h(t) = \int_{\mathbb{C}^n} (f *_3 f^*)(z,t)\, dz$$

then we have

$$\hat{h}(\lambda) = (2\pi)^n |\lambda|^n \|\hat{f}(\lambda)\|_{HS}^2.$$

Now the conditions on f and $\hat{f}$ translate into the conditions

$$|h(t)| \leq Ce^{-\frac{a}{2}t^2}, \ |\hat{h}(\lambda)| \leq Ce^{-2b'\lambda^2}$$

where b' can be chosen so that $ab' > \frac{1}{4}$ or $ab' < \frac{1}{4}$ according as $ab > \frac{1}{4}$ or $ab < \frac{1}{4}$. When $ab > \frac{1}{4}$, Hardy's theorem on $\mathbb{R}$ implies that $h = 0$. This means that $\hat{f}(\lambda) = 0$ and hence $f = 0$ by the Plancherel theorem. If $ab < \frac{1}{4}$ then any function of the form $g(z)h_k(t)$ where h_k are the one dimensional Hermite functions satisfies the hypotheses of the theorem. ∎

    In the above theorem the condition on the Fourier transform imposes very strong restrictions on the function. To see this consider functions of the form $f(z,t) = g(z)h(t)$. Then $\|\hat{f}(\lambda)\|_{HS} \leq Ce^{-b\lambda^2}$ holds if and only if $|\hat{h}(\lambda)| \leq Ce^{-b\lambda^2}$. As we can see from the proof, the above is essentially a theorem for the t variable. We would like to replace the condition on $\hat{f}(\lambda)$ by a weaker condition. We will ignore the t variable completely and prove Hardy's theorem for the Weyl transform. So we need a suitable operator analogue of the conditon $|\hat{f}(\xi)| \leq Ce^{-b|\xi|^2}$.

Our conditions on the operator involve singular numbers. Recall that if T is a compact operator, then TT^* is self adjoint and nonnegative so that its eigenvalues $\lambda_n(T)^2$ form a decreasing sequence. These $\lambda_n(T)$ are called the singular numbers of T and they are invariant under unitary conjugation. If T is an integral operator with kernel $K(x,y)$, then the decay of the singular numbers and the smoothness of the kernel are related. For example $\{\lambda_n(T)\}$ is rapidly decreasing if and only if T is unitarily equivalent to an integral operator whose kernel is in the Schwartz class. We are interested in operators whose singular numbers are exponentially decreasing.

We state our conditions on the operator in terms of the unbounded operator e^{aH}, $a > 0$. These operators are densely defined; their domain consists of all finite linear combinations of Hermite functions. If $f = \sum_{k=0}^m c_k P_k f$ then we define

$$e^{aH} f = \sum_{k=0}^m c_k e^{(2k+n)a} P_k f.$$

We consider operators T for which Te^{aH} is Hilbert-Schmidt for some positive a. This will be the analogue of the condition $|\hat{f}(\xi)| \leq Ce^{-b|\xi|^2}$ for the Fourier transform. Later we will show that the above condition translates into exponential decay of singular numbers of certain Fourier coefficients of T. Actually, when T is diagonalised by the Hermite basis, it is easy to see that this is so. In that case $T = \varphi(H)$ for some function φ and hence

$$Te^{aH} = \sum_{k=0}^\infty \varphi(2k+n)e^{(2k+n)a} P_k.$$

Therefore, $\|Te^{aH}\|_{HS} < \infty$ implies that $|\varphi(2k+n)| \leq Ce^{-(2k+n)a}$ and $\varphi(2k+n)$ are precisely the singular numbers of T.

If f is compactly supported, then $\hat{f}$ cannot have any exponential decay. This follows from the fact that if $\hat{f}$ has any exponential decay, then f extends to a function on $\mathbb{C}^n$ which is holomorphic in a strip. We now prove a result which is the analogue of this property for the Weyl transform.

Theorem 1.6.3 *Let $f \in L^2(\mathbb{C}^n)$ be compactly supported. If*

$$\|W(f)e^{aH}\|_{HS} < \infty$$

for some positive a then $f = 0$.

Proof: We will show that if $W(f)$ verifies the condition of the theorem then f is real analytic. This will immediately prove the theorem. In order to show that f is real analytic, we use an elliptic regularity theorem of Kotake and Narasimhan: Let $P(x, D)$ be a second order elliptic differential operator on $\mathbb{R}^n$ with real analytic coefficients. If a function f satisfies the estimates

$$\|P(x, D)^m f\|_2 \leq M^{m+1}(2m)!$$

for all positive integers m, where M is a constant, then f is real analytic. A proof of this theorem can be found in Narasimhan [44].

Now the function f has the special Hermite expansion

$$f = (2\pi)^{-n} \sum_{k=0}^{\infty} f \times \varphi_k$$

and, since $f \times \varphi_k$ are eigenfunctions of the special Hermite operator L, we have

$$\|L^m f\|_2^2 = (2\pi)^{-2n} \sum_{k=0}^{\infty} (2k + n)^{2m} \|f \times \varphi_k\|_2^2.$$

The hypothesis on $W(f)e^{aH}$ gives us

$$\|W(f)P_k\|_{HS} \leq C e^{-(2k+n)a}$$

and as $W(f)P_k = (2\pi)^{-n} W(f \times \varphi_k)$ in view of the Plancherel theorem, we get $\|f \times \varphi_k\|_2 \leq C e^{-(2k+n)a}$. Consequently the above series giving $\|L^m f\|_2$ converges uniformly and gives the estimate

$$\|L^m f\|_2^2 \leq C \sum_{k=0}^{\infty} (2k + n)^{2m} e^{-2(2k+n)a}.$$

This series can be estimated by the integral $\int_0^\infty t^{2m} e^{-2at}\, dt$ which gives the estimate $\|L^m f\|_2^2 \leq M^{2m+1}(2m)!$ which is more than what we need. Hence by the regularity theorem f is real analytic.   ∎

We also have the following version of the above theorem for the group Fourier transform.

Theorem 1.6.4 *Let $f \in L^2(H^n)$ be compactly supported. Suppose there is a compact set B of $\mathbb{R}^*$ such that for every λ not in B there exists $a(\lambda) > 0$ for which $\hat{f}(\lambda)e^{a(\lambda)H} \in S_2$. Then $f = 0$.*

Proof: Since $\hat{f}(\lambda) \in S_2$, there is a function $f_\lambda \in L^2(\mathbb{C}^n)$ such that $\hat{f}(\lambda) = W(f_\lambda)$. This function can be explicitly calculated as follows: $f_\lambda = |\lambda|^{-n} f^\lambda(\lambda^{-1}x + iy)$. From this it is clear that $f_\lambda(z)$ is compactly supported. As $W(f_\lambda)e^{a(\lambda)H} \in S_2$, by the previous theorem, we get $f_\lambda = 0$ for any λ not in B. This means that $f^\lambda(z) = 0$ for all z and λ not in B which is not possible as $f(z,.)$ is compactly supported. Hence the theorem. ∎

Finally, we state and prove the promised Hardy theorem for the Weyl transform.

Theorem 1.6.5 *Let f be a measurable function which satisfies the estimate $|f(z)| \leq Ce^{-a|z|^2}$. Further assume that $W(f)e^{bH} \in S_2$ for some $b > 0$. Then $f = 0$ whenever $a(\tanh \frac{b}{2}) > \frac{1}{4}$.*

The basic idea of the proof is to reduce it to the case of the Hardy theorem on $\mathbb{C}^n$. As we cannot directly get estimates on $\hat{f}$, we consider the Fourier coefficients f_m of f and get estimtates for $\hat{f}_m$. We will then conclude that $f_m = 0$ for all m which will prove the theorem.

So we look at the multiple Fourier series

$$f(e^{i\theta}z) = \sum f_m(z)e^{-im.\theta}$$

where $f_m(z) = R_{-m}f(z)$ are the $(-m)-$radialisation of f. The exponential decay of f gives the estimates

$$|f_m(z)| \leq Ce^{-a|z|^2}.$$

We will show that for each m the estimates for the Euclidean Fourier transform , $\hat{f}_m$:

$$|\hat{f}_m(z)|^2 \leq Ce^{-2\tanh \frac{b}{2}|z|^2}$$

hold. Then by Hardy's theorem on $\mathbb{C}^n$, we get $f_m = 0$ whenever $a(\tanh \frac{b}{2}) > \frac{1}{4}$.

In order to estimate the Fourier transforms of f_m, we need to get good estimates on their special Hermite coefficients. We need to recall a few facts about the so-called metaplectic representation. For each $\sigma \in U(n)$, the group of $n \times n$ unitary matrices, we have an automorphism $(z,t) \to (\sigma z, t)$ of the Heisenberg group. That this is an automorphism follows from the fact that $U(n)$ preserves the symplectic form $[z, w] = Im(z.\bar{w})$. If ρ is a representation of H^n then $\rho_\sigma(z,t) = \rho(\sigma z, t)$ is another representation agreeing with ρ at the center, and so by the Stone-von Neumann theorem, ρ_σ is unitarily equivalent to ρ.

This reasoning applied to the Schrödinger representation $\pi_1(z,t)$ shows that for each $\sigma \in U(n)$, there is a unitary operator $\mu(\sigma)$ such that

$$\pi(\sigma z) = \mu(\sigma)\pi(z)\mu(\sigma)^*. \qquad (1.6.46)$$

This operator valued function μ can be chosen in such a way that it defines a unitary representation of the double cover of the symplectic group and is called the metaplectic representation. We will not go into a systematic study of the metaplectic representation as it requires a considerable amount of background material on the symplectic group. However, we need to use many basic properties of this representation. A nice introduction to this representation is given in Folland [26] and we will use results from there without further comments. For the proof of Hardy's theorem, we need the following fact: for $\sigma \in T(n)$ the operators $\mu(\sigma)$ commute with the Hermite operator H.

We now calculate the Weyl transform of f_m. From the definition of f_m, it follows that

$$W(f_m) = (2\pi)^{-n} \int_{\mathbb{C}^n} \int_{T(n)} f(e^{i\theta}z)e^{im.\theta}\pi(z)\, d\theta\, dz.$$

Making a change of variables and noting that

$$\pi(e^{-i\theta}z) = \mu(e^{-i\theta})\pi(z)\mu(e^{i\theta}),$$

we get the formula

$$W(f_m) = (2\pi)^{-n} \int_{T(n)} \mu(e^{-i\theta})W(f)\mu(e^{i\theta})e^{im.\theta}\, d\theta \qquad (1.6.47)$$

which may be called the mth Fourier coefficient of the operator $W(f)$. Since $\mu(e^{-i\theta})$ commutes with e^{bH}, we get

$$W(f_m)e^{bH} = (2\pi)^{-n} \int_{T(n)} \mu(e^{-i\theta})W(f)e^{bH}\mu(e^{i\theta})e^{im.\theta}\, d\theta.$$

From this formula it follows that $W(f_m)e^{bH} \in S_2$ whenever $W(f)e^{bH} \in S_2$.

We now show that the condition $W(f_m)e^{bH} \in S_2$ translates into the exponential decay of the singular numbers of $W(f_m)$. To see this we calculate $W(f_m)$ by expanding f_m in terms of $\bar{\Phi}_{\alpha,\beta}$. As f_m is $-m$-homogeneous and $\bar{\Phi}_{\alpha,\beta}$ is $(\alpha - \beta)$-homogeneous, we get

$$f_m = \sum_{\beta}(f_m, \bar{\Phi}_{\beta,\beta+m})\bar{\Phi}_{\beta,\beta+m}.$$

From the properties of the Fourier-Wigner transform it follows that

$$W(f_m)\varphi = \sum_{\beta}(f_m, \bar{\Phi}_{\beta,\beta+m})(\varphi, \Phi_\beta)\Phi_{\beta+m}.$$

Thus the singular numbers of $W(f_m)$ are given by $\lambda_\beta^m = |(f_m, \bar{\Phi}_{\beta,\beta+m})|$. Since Φ_β are eigenfunctions of H, we get

$$W(f_m)e^{bH}\varphi = \sum_{\beta}(f_m, \bar{\Phi}_{\beta,\beta+m})e^{(2|\beta|+n)b}(\varphi, \Phi_\beta)\Phi_{\beta+m}.$$

and the condition $W(f_m)e^{bH} \in S_2$ gives us the exponential decay

$$|(f_m, \bar{\Phi}_{\beta,\beta+m})| \le Ce^{-(2|\beta|+n)b}.$$

Using these conditions we will estimate the Fourier transform of f_m.

The symplectic Fourier transform of a function f defined on $\mathbb{C}^n$ is given by

$$\mathcal{F}_s f(z) = 2^{-n}\int_{\mathbb{C}^n} f(w)e^{-\frac{i}{2}Im(z.\bar{w})}\,dw.$$

In terms of the Euclidean Fourier transform on $\mathbb{R}^{2n}$, we have the relation

$$\mathcal{F}_s f(z) = \pi^n \hat{f}(\frac{y}{2}, -\frac{x}{2}).$$

Observe that we can write

$$\mathcal{F}_s f(z) = 2^{-n}\int_{\mathbb{C}^n} f(z-w)e^{\frac{i}{2}Im(z.\bar{w})}\,dw$$

so that $\mathcal{F}_s f(z) = 2^{-n}f \times 1$ where 1 is the constant function 1. We claim that

$$\mathcal{F}_s\Phi_{\beta,\beta+m} = (4\pi)^{-n}(-1)^{|\beta+m|}\Phi_{\beta,\beta+m}. \qquad (1.6.48)$$

The special Hermite functions $\Phi_{\mu,\mu}$ can be expressed in terms of multiple Laguerre functions and they satisfy the generating function identity

$$\sum_{\mu} r^{|\mu|}\Phi_{\mu,\mu} = (1-r)^{-n}e^{-\frac{1}{4}(\frac{1+r}{1-r})|z|^2}.$$

Calculating the Fourier transform on both sides we have

$$\sum_{\mu} r^{|\mu|}\hat{\Phi}_{\mu,\mu} = \pi^{-n}(1+r)^{-n}e^{-(\frac{1-r}{1+r})|z|^2}$$

from which we obtain the identity

$$\sum_{\mu} r^{|\mu|} \mathcal{F}_s \Phi_{\mu,\mu} = \sum_{\mu} (-r)^{|\mu|} \Phi_{\mu,\mu}.$$

This means that $\mathcal{F}_s \Phi_{\mu,\mu} = (-1)^{|\mu|} \Phi_{\mu,\mu}$ and expanding 1 as

$$1 = (2\pi)^{-n} \sum_{\mu} 1 \times \Phi_{\mu,\mu} = (2\pi)^{-n} \sum_{\mu} (-1)^{|\mu|} \Phi_{\mu,\mu}$$

we get

$$\mathcal{F}_s \Phi_{\alpha,\alpha+m} = (4\pi)^{-n} \sum_{\mu} (-1)^{|\mu|} \Phi_{\alpha,\alpha+m} \times \Phi_{\mu,\mu}.$$

In view of the orthogonality property (4.13) the above reduces to

$$\mathcal{F}_s \Phi_{\alpha,\alpha+m} = (4\pi)^{-n} (-1)^{|\alpha+m|} \Phi_{\alpha,\alpha+m}$$

and this proves the claim.

Therefore, we have the expansion

$$\mathcal{F}_s \bar{f}_m(z) = (4\pi)^{-n} \sum_{\alpha} \lambda_{\alpha}^m (-1)^{|\alpha+m|} \Phi_{\alpha,\alpha+m}.$$

The functions $\Phi_{\alpha,\alpha+m}$ are expressible as products of one dimensional Laguerre functions and so without loss of generality we can assume $n = 1$ and m is a nonnegative integer. Then we have

$$|\mathcal{F}_s \bar{f}_m(z)| \leq (4\pi)^{-1} \sum_{k=0}^{\infty} |\lambda_k^m| |\Phi_{\alpha,\alpha+m}(z)|$$

which by Cauchy-Schwarz inequality and the estimates on λ_k^m gives us

$$|\mathcal{F}_s \bar{f}_m(z)|^2 \leq C \sum_{k=0}^{\infty} e^{-(2k+1)b} |\Phi_{k,k+m}(z)|^2.$$

Now, from Proposition 1.4.2,

$$\Phi_{k,k+m}(z) = (2\pi)^{-\frac{1}{2}} \left(\frac{-i}{\sqrt{2}}\right)^m \left\{\frac{k!}{(k+m)!}\right\}^{\frac{1}{2}} z^m L_k^m\left(\frac{1}{2}|z|^2\right) e^{-\frac{1}{4}|z|^2}$$

and the Laguerre functions satisfy the generating function identity (see Szego [72])

$$\sum_{k=0}^{\infty} \frac{\Gamma(k+1)}{(k+m-1)} (L_k^m(s^2))^2 e^{-\frac{1}{2}s^2} r^k$$

$$= (1-r)^{-1}(-s^4 r)^{-\frac{m}{2}} e^{-\frac{1+r}{1-r}s^2} J_m\left(\frac{2is^2\sqrt{r}}{1-r}\right)$$

where J_m is the Bessel function of first kind.

Taking $r = e^{-2b}$ in the above we get

$$\sum_{k=0}^{\infty} e^{-(2k+1)b}|\Phi_{k,k+m}(z)|^2$$

$$= C|z|^{-2m} e^{-\frac{1}{2}\frac{1+e^{-2b}}{1+e^{-2b}}|z|^2} J_m\left(\frac{i|z|^2 e^{-b}}{1-e^{-2b}}\right).$$

The Bessel function $J_m(is)$ satisfies the estimate

$$|J_m(is)| \leq Cs^{-\frac{1}{2}}e^s$$

as $s \to \infty$. Using this in the above we conclude that

$$\sum_{k=0}^{\infty} e^{-(2k+1)b}|\Phi_{k,k+m}(z)|^2 \leq Ce^{-\frac{1}{2}\tanh(\frac{b}{2})|z|^2}.$$

Therefore, we have

$$|\mathcal{F}_s\tilde{f}_m(z)| \leq Ce^{-\frac{1}{4}\tanh(\frac{b}{2})|z|^2}$$

and consequently

$$|\hat{f}_m(z)| \leq Ce^{-\tanh(\frac{b}{2})|z|^2}.$$

Now we appeal to Hardy's theorem for $\mathbb{C}^n$ to conclude that $f_m = 0$ whenever $a(\tanh\frac{b}{2}) > \frac{1}{4}$. This completes the proof of the theorem. ∎

1.7   Notes and references

For a general introduction to the Heisenberg group and its representations, we refer to the books of Folland [26], Taylor [73], Stein [65] and to the papers of Geller [30],[31]. For a nice introduction to the importance of Heisenberg group in analysis, we refer to the paper of Howe [34]. For various properties of Hermite and Laguerre functions, we refer to Szego [72] and for the harmonic analysis of these expansions, refer to the monograph [84] of the author. The Hausdorff-Young inequality is proved in Kunze[37] and Peetre-Sparr [52]. An account of the classical Paley-Wiener theorems with applications to partial differential equations can be found in Rudin [58]. The Paley-Wiener theorems for the

Heisenberg group are established in [85],and [89]. See also the paper of Ando [2]. Hardy's theorem for the real line can be found in Dym-McKean [19]. For analogues of the Hardy's theorem on the Heisenberg group we refer to Sitaram et al [60] and the unpublished manuscript of the author [89].

Chapter 2

ANALYSIS OF THE SUBLAPLACIAN

In this chapter we study the spectral theory of the sublaplacian as developed by Strichartz. We obtain an Abel summability result for expansions in terms of the eigenfunctions of the sublaplacian. For the spectral projection operators, we establish some restriction theorems. Using the restriction theorem, we study the Bochner-Riesz means for the sublaplacian. We also develop a Littlewood-Paley-Stein theory for the sublaplacian and prove a multiplier theorem for the Fourier transform.

2.1 Spectral theory of the sublaplacian

In this section we define the sublaplacian and study its spectral properties. This operator, denoted by $\mathcal{L}$, is the counterpart of the Laplacian Δ on $\mathbb{R}^n$. Recall that Δ is characterised by the following properties: (i) it is invariant under translations and rotations, (ii) it is homogeneous of degree two.(We say that an operator T is homogeneous of degree α if it satisfies

$$T(f(rx)) = r^\alpha (Tf)(rx)$$

for all $r > 0$ and $x \in \mathbb{R}^n$.) On the Heisenberg group we have the left translations L_g defined for $g \in H^n$ by

$$L_g f(h) = f(g^{-1}h), h \in H^n.$$

We also have the rotations

$$R_\sigma f(z,t) = f(\sigma z, t), \ \sigma \in U(n).$$

The rotations $R_\sigma(z,t) = (\sigma z, t)$ are automorphisms of the Heisenberg group. Instead of the usual dilations $x \to rx$, we have the nonisotropic dilations $\delta_r(z,t) = (rz, r^2t)$. It is clear that δ_r are also automorphisms of the Heisenberg group.

Given a differential operator P on H^n we say that it is left invariant if it commutes with L_g for all $g \in H^n$; it is said to be rotation invariant if it commutes with R_σ for all $\sigma \in U(n)$. It is said to be homogeneous of degree a if $P(f(\delta_r g)) = r^a Pf(\delta_r g)$. It can be shown that up to a constant multiple, there is a unique left invariant, rotation invariant differential operator that is homogeneous of degree 2. This unique operator is called the sublaplacian or Kohn-Laplacian on the Heisenberg group and enjoys many properties satisfied by the Laplacian on $\mathbb{R}^n$. A remarkable fundamental solution for this operator was found by Folland [23]. Though the sublaplacian fails to be elliptic, it satisfies certain estimates known as subelliptic estimates; see Folland-Kohn [24]. This operator is closely related to the $\bar\partial$−Neumann problem on the Siegel upper half space. We refer to Chapter XIII of Stein [65] for details of this connection.

The sublaplacian $\mathcal{L}$ is explicitly given by

$$\mathcal{L} = -\sum_{j=1}^{n}(X_j^2 + Y_j^2) \qquad\qquad (2.1.1)$$

where X_j and Y_j are the left invariant vector fields introduced in Section 1.1 . An explicit calculation shows that

$$\mathcal{L} = -\Delta_z - \frac{1}{4}|z|^2\partial_t^2 + N\partial_t$$

where Δ_z is the Laplacian on $\mathbb{C}^n$ and

$$N = \sum_{j=1}^{n}(x_j\frac{\partial}{\partial y_j} - y_j\frac{\partial}{\partial x_j})$$

is the rotation operator. From this formula it follows that $\mathcal{L}$ has all the properties mentioned above. We will now investigate the spectrum of this operator. Actually we treat the joint spectrum of the operators $\mathcal{L}$ and $T = \partial_t$. Since these two operators commute, the joint spectrum is well defined. We will explicitly write down the eigenfunctions of these operators in terms of special Hermite functions.

We look for eigenfunctions that arise as matrix components of irreducible unitary representations. Recall that if X is a left invariant vector field on H^n and π is a representation, then $\pi(X)$ is defined to be a skew adjoint unbounded operator acting on the C^∞ vectors of the representation. Let π be a representation of H^n on a Hilbert space $\mathcal{H}$ and consider $(\pi(g)u, v)$ where $g \in H^n$ and $u, v \in \mathcal{H}$. If A is any element of the universal enveloping algebra of left invariant differential operators on the Heisenberg group, then an easy calculation shows that

$$A(\pi(g)u, v) = (\pi(g)\pi(A)u, v).$$

When A is in the centre of the universal enveloping algebra, $\pi(A)$ must be a constant multiple of the identity whenever π is irreducible, and therefore if $\pi(A) = \lambda I$ then $A(\pi(g)u, v) = \lambda(\pi(g)u, v)$; therefore any matrix component $(\pi(g)u, v)$ is an eigenfunction of A. On the other hand when A is not in the centre of the universal enveloping algebra, then the matrix component $(\pi(g)u, v)$ will be an eigenfunction of A whenever u is an eigenfunction of $\pi(A)$.

When A is the sublaplacian and π is one of the Schrödinger representations, it is easy to calculate $\pi(A)$. In fact, a direct calculation shows that

$$\pi_\lambda(X_j)\varphi(\xi) = -i\lambda\xi_j\varphi(\xi)$$

and

$$\pi_\lambda(Y_j)\varphi(\xi) = \frac{\partial}{\partial\xi_j}\varphi(\xi)$$

and consequently

$$\pi_\lambda(\mathcal{L}) = -\Delta + \lambda^2|\xi|^2 = H(\lambda)$$

is the scaled Hermite operator. The functions

$$\Phi_\alpha^\lambda(\xi) = |\lambda|^{\frac{n}{4}}\Phi_\alpha(|\lambda|^{\frac{1}{2}}\xi)$$

are eigenfunctions of the operator $H(\lambda)$ with eigenvalues $(2|\alpha| + n)|\lambda|$. Thus, taking $u = \Phi_\alpha^\lambda$ and $v = \Phi_\beta^\lambda$, we observe that the functions $(\pi_\lambda(z, t)\Phi_\alpha^\lambda, \Phi_\beta^\lambda)$ are eigenfunctions of the sublaplacian:

$$\mathcal{L}(\pi_\lambda(z, t)\Phi_\alpha^\lambda, \Phi_\beta^\lambda) = (2|\alpha| + n)|\lambda|(\pi_\lambda(z, t)\Phi_\alpha^\lambda, \Phi_\beta^\lambda).$$

These matrix components can be expressed in terms of special Hermite functions. For $\lambda > 0$, we have

$$(\pi_\lambda(z, t)\Phi_\alpha^\lambda, \Phi_\beta^\lambda) = (2\pi)^{\frac{n}{2}}e^{i\lambda t}\Phi_{\alpha,\beta}(\sqrt{\lambda}z)$$

and when $\lambda < 0$, we have

$$(\pi_\lambda(z, t)\Phi_\alpha^\lambda, \Phi_\beta^\lambda) = (2\pi)^{\frac{n}{2}}e^{i\lambda t}\bar{\Phi}_{\alpha,\beta}(\sqrt{-\lambda}z).$$

Note that these functions are also eigenfunctions of iT with eigenvalues $-\lambda$.

From the explicit form it is clear that the sublaplacian is polyradial, that is, it commutes with the action of $T(n)$. The eigenfunctions that are polyradial are given by the Laguerre functions $e^{i\lambda t}\Phi_{\alpha,\alpha}(\sqrt{|\lambda|}z)$ and as

$$\varphi_k(z) = (2\pi)^{\frac{n}{2}} \sum_{|\alpha|=k} \Phi_{\alpha,\alpha}(z)$$

functions of the form $e^{i\lambda t}\varphi_k(\sqrt{|\lambda|}z)$ are radial eigenfunctions of $\mathcal{L}$. For each $\lambda \in \mathbb{R}^*$ we define

$$e_k^\lambda(z,t) = e^{i\lambda t}\varphi_k^\lambda(z) = e^{i\lambda t}\varphi_k(\sqrt{|\lambda|}z).$$

Then it is clear that $e_k^\lambda(z,t)$ are joint eigenfunctions of $\mathcal{L}$ and T the joint spectrum being the set

$$S = \{(\lambda, (2k+n)|\lambda|) : \lambda \in R^*, k \in N\}.$$

This set is called the Heisenberg fan.

Since $\mathcal{L}$ is left invariant, $\mathcal{L}(f * g) = f * \mathcal{L}g$ and hence $f * e_k^\lambda$ are eigenfunctions of the sublaplacian with eigenvalues $(2k+n)|\lambda|$. This operator taking f into $f * e_k^\lambda$ is nothing but the spectral projection associated to the ray

$$R_k = \{(\lambda, (2k+n)|\lambda|) : \lambda \in R^*\}$$

of the spectrum. We now show that we can expand functions f in terms of their spectral projections.

Theorem 2.1.1 *For $f \in L^2(H^n)$ we have the expansion*

$$f(z,t) = \sum_{k=0}^{\infty} \int_{-\infty}^{\infty} f * e_k^\lambda(z,t)\, d\mu(\lambda)$$

where $d\mu(\lambda) = (2\pi)^{-n-1}|\lambda|^n d\lambda$ is the Plancherel measure for the Heisenberg group.

Proof: Recall the Fourier inversion formula

$$f(z,t) = \int_{-\infty}^{\infty} tr(\pi_\lambda(z,t)^* \hat{f}(\lambda))\, d\mu(\lambda).$$

Assuming $\lambda > 0$ let us calculate the above trace by using the orthonormal basis Φ_α^λ. We have

$$tr(\pi_\lambda(z,t)^*\hat{f}(\lambda)) = e^{-i\lambda t} \sum_\alpha \left(\pi_\lambda(z)^*\hat{f}(\lambda)\Phi_\alpha^\lambda, \Phi_\alpha^\lambda\right).$$

Writing the definition of $\hat{f}(\lambda)$, the above equals

$$e^{-i\lambda t} \sum_\alpha \int_{\mathbb{C}^n} f^\lambda(w)(\pi_\lambda(w)\Phi_\alpha^\lambda, \pi_\lambda(z)\Phi_\alpha^\lambda)\, dw.$$

Now we expand $\pi_\lambda(w)\Phi_\alpha^\lambda$ in terms of Φ_β^λ. We get the series

$$\pi_\lambda(w)\Phi_\alpha^\lambda(\xi) = \sum_\beta (\pi_\lambda(w)\Phi_\alpha^\lambda, \Phi_\beta^\lambda)\Phi_\beta^\lambda(\xi)$$

and from this we get

$$(\pi_\lambda(w)\Phi_\alpha^\lambda, \pi_\lambda(z)\Phi_\alpha^\lambda) = \sum_\beta (\pi_\lambda(w)\Phi_\alpha^\lambda, \Phi_\beta^\lambda)(\Phi_\beta^\lambda, \pi_\lambda(z)\Phi_\alpha^\lambda).$$

We remark that

$$(\pi_\lambda(w)\Phi_\alpha^\lambda, \Phi_\beta^\lambda) = (2\pi)^{\frac{n}{2}}\Phi_{\alpha,\beta}(\sqrt{\lambda}w),$$

and hence

$$tr(\pi_\lambda(z,t)^*\hat{f}(\lambda))$$

$$= (2\pi)^n e^{-i\lambda t} \sum_\alpha \sum_\beta \left(\int_{\mathbb{C}^n} f^\lambda(w)\Phi_{\alpha,\beta}(\sqrt{\lambda}w)\, dw\right) \bar{\Phi}_{\alpha,\beta}(\sqrt{\lambda}z).$$

The above is the special Hermite expansion of $f^\lambda(w)$ which can be put in the form

$$\sum_\alpha \sum_\beta \left(\int_{\mathbb{C}^n} f^\lambda(w)\Phi_{\alpha,\beta}(\sqrt{\lambda}w)\, dw\right) \bar{\Phi}_{\alpha,\beta}(\sqrt{\lambda}z)$$

$$= (2\pi)^{-n} \sum_{k=0}^\infty f^\lambda *_\lambda \varphi_k^\lambda(z).$$

Putting this back and recalling the definition of e_k^λ, we have

$$tr(\pi_\lambda(z,t)^*\hat{f}(\lambda)) = \sum_{k=0}^\infty f * e_k^\lambda(z,t) \qquad (2.1.2)$$

and we can do the calculation for $\lambda < 0$ as well. The Fourier inversion formula then gives us

$$f(z,t) = \sum_{k=0}^{\infty} \int_{-\infty}^{\infty} f * e_k^{\lambda}(z,t) \, d\mu(\lambda)$$

and this completes the proof of the theorem.   ∎

We may consider the above expansion as the analogue of the Peter-Weyl theorem for the Heisenberg group. We now establish the Plancherel theorem for the above expansion.

Theorem 2.1.2 *For $f \in L^2(H^n)$ we have*

$$\|f\|_2^2 = (2\pi)^{-2n-1} \sum_{k=0}^{\infty} \int_{-\infty}^{\infty} \int_{\mathbb{C}^n} |f * e_k^{\lambda}(z,0)|^2 \lambda^{2n} \, dz \, d\lambda.$$

Proof: We use formula (2.1.2) to calculate the Hilbert-Schmidt operator norm of $\hat{f}(\lambda)$. For $\lambda > 0$ from the definition of $\hat{f}(\lambda)$ we have

$$tr(\hat{f}(\lambda)^* \hat{f}(\lambda)) = \int_{\mathbb{C}^n} \bar{f}(z,t) tr(\pi_{\lambda}(z,t)^* \hat{f}(\lambda)) \, dz \, dt.$$

In view of (2.1.2) the above equals

$$\int_{\mathbb{C}^n} \left(\sum_{k=0}^{\infty} f^{\lambda} *_{\lambda} \varphi_k^{\lambda}(z) \right) \bar{f}^{\lambda}(z) \, dz.$$

We then expand f^{λ} as

$$f^{\lambda}(z) = (2\pi)^{-n} \lambda^n \sum_{k=0}^{\infty} f^{\lambda} *_{\lambda} \varphi_k^{\lambda}(z),$$

and using the orthogonality properties of the special Hermite projections we get

$$tr(\hat{f}(\lambda)^* \hat{f}(\lambda)) = (2\pi)^{-n} \lambda^n \sum_{k=0}^{\infty} \int_{\mathbb{C}^n} |f^{\lambda} *_{\lambda} \varphi_k^{\lambda}(z)|^2 dz.$$

Performing a similar calculation for $\lambda < 0$ and applying the Plancherel theorem, we get

$$\|f\|_2^2 = (2\pi)^{-2n-1} \sum_{k=0}^{\infty} \int_{-\infty}^{\infty} \int_{\mathbb{C}^n} |f^{\lambda} * \varphi_k^{\lambda}(z)|^2 \lambda^{2n} \, dz \, d\lambda.$$

Since

$$f^\lambda *_\lambda \varphi_k^\lambda(z) = f * e_k^\lambda(z, 0)$$

this proves the theorem. ∎

Later in Chapter 3 we will give a representation theoretic interpretation of the above expansion in terms of the Heisenberg motion group. In the next section we study the above expansion for L^p functions.

2.2 Spectral decomposition for L^p functions

In Theorem 2.1.1 we obtained the decomposition under the assumption that $f \in L^2(H^n)$. In this section we wish to study the above expansion when $f \in L^p(H^n)$. A priori, it is not even clear why the above expansion makes sense. Once we show that the expansion is well defined, there is then the problem of investigating in what sense the resulting series converges. To deal with these questions we have to look more closely the spectral projection operators associated to the ray R_k. These are the operators taking f into

$$\int_{-\infty}^{\infty} f * e_k^\lambda(z, t) \, d\mu(\lambda).$$

To define these projections for $f \in L^p(H^n)$ we proceed as follows.
Let $f \in L^p(H^n)$. Then for each $a > 0$ the convolution

$$f * \int_{-a}^{a} e_k^\lambda(z, t) |\lambda|^n \, d\lambda$$

is well defined. To see this we observe that the function

$$E_k^a(z, t) = \int_{-a}^{a} e_k^\lambda(z, t) |\lambda|^n \, d\lambda$$

satisfies the estimate

$$\|E_k^a\|_q^q \leq \int_{\mathbb{C}^n} \left(\int_{-a}^{a} |\varphi_k^\lambda(z)|^q |\lambda|^{np} \, d\lambda \right)^{\frac{q}{p}} dz$$

where $1 \leq p \leq 2$ and q is the conjugate exponent of p. This is just a consequence of the Hausdorff-Young inequality for the Fourier transform in the t variable. By an application of Minkowski's integral inequality, we get

$$\|E_k^a\|_q \leq \|\varphi_k\|_q \left(\int_{-a}^{a} |\lambda|^{np} \, d\lambda \right)^{\frac{1}{p}}$$

and therefore the convolution $f * E_k^a$ is well defined whenever $f \in L^p(H^n)$, $1 \leq p \leq 2$. By Fubini's theorem we have

$$f * E_k^a(z,t) = \int_{-a}^{a} f * e_k^\lambda(z,t)|\lambda|^n \, d\lambda.$$

Suppose now that f is a Schwartz class function such that $f^\lambda(z)$ is compactly supported as a function of λ, say in $|\lambda| \leq b$. Then the convolution $f * E_k^a$ is independent of a as long as $a > b$. To see this we have

$$f * E_k^a(z,t) = \int_{-a}^{a} f * e_k^\lambda(z,t)|\lambda|^n \, d\lambda$$

and as

$$f * e_k^\lambda(z,t) = e^{-i\lambda t} f^\lambda *_\lambda \varphi_k^\lambda(z)$$

is supported in $|\lambda| \leq b$ we get

$$f * E_k^a(z,t) = \int_{-b}^{b} f * e_k^\lambda(z,t)|\lambda|^n \, d\lambda.$$

Thus if f is a Schwartz class function with the property that $f^\lambda(z)$ is compactly supported in λ, then the convolution

$$\int_{-\infty}^{\infty} f * e_k^\lambda(z,t)|\lambda|^n \, d\lambda$$

is well defined and is equal to

$$f * \int_{-\infty}^{\infty} e_k^\lambda(z,t)|\lambda|^n \, d\lambda.$$

This latter integral must be interpreted in the principal value sense.

We now proceed to calculate the kernel $\int_{-\infty}^{\infty} e_k^\lambda(z,t)|\lambda|^n \, d\lambda$ explicitly. By writing

$$\int_{-\infty}^{\infty} e_k^\lambda(z,t)|\lambda|^n \, d\lambda$$

$$= \int_{0}^{\infty} e_k^\lambda(z,t)\lambda^n \, d\lambda + \int_{0}^{\infty} e_k^{-\lambda}(z,t)\lambda^n \, d\lambda$$

and observing that

$$e_k^{-\lambda}(z,t) = e_k^\lambda(z,-t)$$

it is enough to calculate one integral in the above equation.

Proposition 2.2.1 *There is an integrable function $F_k(s)$ such that*

$$\int_0^\infty e_k^\lambda(z,t)\lambda^n \, d\lambda = |z|^{-2n-2} F_k(t|z|^{-2})$$

for all $z \neq 0$.

Proof: By making a change of variable, it is easy to see that

$$\int_0^\infty e_k^\lambda(z,t)\lambda^n \, d\lambda = |z|^{-2n-2} F_k(t|z|^{-2})$$

where the function $F_k(t)$ is defined by

$$F_k(t) = \int_0^\infty e^{-its}\varphi_k(\sqrt{s})s^n \, ds.$$

To compute the function $F_k(t)$ we use the generating function

$$\sum_{k=0}^\infty r^k L_k^{n-1}(\tfrac{1}{2}s)e^{-\frac{1}{4}s} = (1-r)^{-n}e^{-\frac{1}{4}(\frac{1+r}{1-r})s}.$$

This gives us the formula

$$\sum_{k=0}^\infty r^k F_k(t) = (1-r)^{-n}\int_0^\infty e^{-its}e^{-\frac{1}{4}(\frac{1+r}{1-r})s}s^n \, ds.$$

The integral on the right hand side equals

$$\int_0^\infty e^{-(A+it)s}s^n \, ds = \Gamma(n+1)(A+it)^{-n-1}$$

where we have written $4A = \frac{1+r}{1-r}$. Therefore,

$$F_k(t) = 4^{n+1}\frac{\Gamma(n+1)}{\Gamma(k+1)}(\frac{d}{dr})^k\{\frac{1+r}{1-r} + 4it\}^{-n-1}|_{r=0}.$$

A simple calculation shows that $F_k(t)$ is equal to

$$(-1)^k\frac{\Gamma(n+k+1)}{\Gamma(k+1)}\{1 + \frac{k}{n+k}\frac{1+4it}{1-4it}\}\frac{(1+4it)^k}{(1-4it)^{n+k+1}}$$

which is clearly an integrable function. ∎

From the proposition it is clear that the kernels

$$\int_{-\infty}^\infty e_k^\lambda(z,t)|\lambda|^n \, d\lambda$$

are homogeneous of degree $(-2n - 2)$ with respect to the Heisenberg dilations $\delta_r(z, t) = (rz, r^2 t)$. Thus the convolution

$$f * \int_{-\infty}^{\infty} e_k^\lambda(z, t)|\lambda|^n \, d\lambda$$

has to be interpreted in the principal value sense and so we need to verify that the kernels satisfy appropriate cancellation conditions in order to show that they are Calderon-Zygmund kernels. We refer to Chapter XII of Stein [65] for a discussion on singular integrals on the Heisenberg group. For a kernel $K(z, t)$ which is homogeneous of degree $(-2n - 2)$, the traditional cancellation condition is equivalent to

$$\int_{\mathbb{C}^n} (K(z, 1) + K(z, -1)) \, dz = 0$$

and it is easy to show that our kernels $|z|^{-2n-2} F_k(t|z|^{-2})$ satisfy this condition. In fact, in polar coordinates, the integral

$$\int_{\mathbb{C}^n} |z|^{-2n-2} \left(F_k(|z|^{-2}) + F_k(-|z|^{-2}) \right) \, dz$$

transforms into

$$\int_0^\infty t^{-3} \left(F_k(t^{-2}) + F_k(-t^{-2}) \right) \, dt = C \int_{-\infty}^\infty F_k(t) \, dt.$$

This integral can be evaluated using the calculus of residues. The function $F_k(z)$ has single pole in the lower half plane. We can choose a contour in the upper half plane to get

$$\int_{-\infty}^\infty F_k(t) \, dt = 0.$$

This proves that the kernels have the required cancellation property. Finally, define the kernels $G_k(z, t)$ by

$$G_k(z, t) = (2\pi)^{-n-1}|z|^{-2n-2} \left(F_k(t|z|^{-2}) + F_k(-t|z|^{-2}) \right).$$

Let $p.v. f * G_k(z, t)$ stand for the principal value convolution

$$p.v. f * G_k(z, t)$$

$$= \lim_{\epsilon \to 0} \int_{|s|>\epsilon} \int_{\mathbb{C}^n} f((z, t)(-w, -s)) G_k(w, s) \, dw \, ds.$$

Let $R_k f$ be the spectral projection associated to the kth ray

$$\{(\lambda, (2k+n)|\lambda|) : \lambda \in \mathbb{R}^*\}$$

of the Heisenberg fan. Formally, we have

$$R_k f(z,t) = \int_{-\infty}^{\infty} f * e_k^{\lambda}(z,t) \, d\mu(\lambda).$$

When the function f is such that f^{λ} is compactly supported in λ, we have

$$R_k f(z,t) = \lim_{a \to \infty} f * \int_{-a}^{a} e_k^{\lambda}(z,t) \, d\mu(\lambda).$$

For general functions interchanging the order of integration and convolution in the definition of $R_k f$ is a delicate matter. We have to introduce a summability factor inside the integral and define $R_k f$ by a limiting argument. What we have in mind is the Abel means

$$f * \int_{-\infty}^{\infty} e_k^{\lambda}(z,t) e^{-s|\lambda|} \, d\mu(\lambda).$$

Regarding the convergence of these means we have the following result.

Theorem 2.2.2 *For $f \in L^2(H^n)$ we have*

$$R_k f(z,t) = \lim_{s \to 0} f * \int_{-\infty}^{\infty} e_k^{\lambda}(z,t) e^{-s|\lambda|} \, d\mu(\lambda),$$

the limit existing in L^2 norm. We also have

$$\lim_{s \to 0} f * \int_{0}^{\infty} e_k^{\lambda}(z,t) e^{-s\lambda} \, d\mu(\lambda)$$

$$= c_{k,n} f + p.v. f * \int_{0}^{\infty} e_k^{\lambda}(z,t) \, d\mu(\lambda)$$

where $c_{k,n}$ is a constant. A similar result holds for the integral taken from $-\infty$ to 0. Thus

$$R_k f(z,t) = 2c_{k,n} f + p.v. f * G_k(z,t)$$

where G_k is the Calderon-Zygmund kernel defined above.

Proof: As we have done in Proposition 2.2.1, we can explicitly calculate the kernels with summability factors. We can show that

$$\int_0^\infty e_k^\lambda(z,t) e^{-s\lambda} |\lambda|^n \, d\lambda = |z|^{-2n-2} F_k((t+is)|z|^{-2})$$

where F_k is as in Proposition 2.2.1. From this explicit formula it is clear that the kernel is the sum of an L^1 function and an L^2 function so that convolution with an L^2 function makes sense; and we can change the order of integration and convolution. The first assertion of the theorem then follows from spectral theory.

Assume that f is a test function, and compare the distribution

$$f \rightarrow \lim_{s\to 0} f * \left(\int_0^\infty e_k^\lambda e^{-\lambda s} \, d\mu(\lambda) \right)(0)$$

with

$$f \rightarrow p.v. f * \left(\int_0^\infty e_k^\lambda \, d\mu(\lambda) \right)(0).$$

The difference between them must be a distribution supported at the origin because the kernel with the summability factor converges to

$$|z|^{-2n-2} F_k(t|z|^{-2})$$

uniformly on compact subsets not containing the origin. By homogeneity arguments. We conclude that the difference is a constant times the delta function. The exact value of the constant can be calculated, for which we refer to Strichartz [71]. ∎

Since the operators R_k are spectral projections, they are bounded on L^2 and since they are Calderon-Zygmund singular integral operators, they are also bounded on $L^p(H^n)$ for $1 < p < \infty$. Therefore, each $f \in L^p(H^n)$ has the formal expansion

$$f(z,t) = \sum_{k=0}^\infty R_k f(z,t).$$

In order to study the convergence properties of the above expansion, we need to estimate the norms of R_k as operators on $L^p(H^n)$.

Before proceeding to get estimates for the norms of R_k as operators on $L^p(H^n)$, we pause to make the following observation. Suppose

$m(\lambda)$ is a function on the real line which defines a bounded L^p multiplier for the Euclidean Fourier transform on $\mathbb{R}$. Then the operators

$$T_m f(z,t) = \int_{-\infty}^{\infty} m(\lambda) f * e_k^{\lambda}(z,t) |\lambda|^n \, d\lambda$$

are bounded on $L^p(H^n)$. To see this we observe that

$$T_m f(z,t) = \int_{-\infty}^{\infty} m(\lambda) e^{-i\lambda t} f^{\lambda} *_{\lambda} \varphi_k^{\lambda}(z,t) |\lambda|^n \, d\lambda$$

so that we can write

$$T_m f(z,t) = R_k M f(z,t)$$

where $(Mf)^{\lambda} = m(\lambda) f^{\lambda}$ is the Euclidean Fourier multiplier, and from this follows our claim.

In particular, operators of the form

$$T_a^b f = \int_a^b f * e_k^{\lambda}(z,t) \, d\mu(\lambda)$$

are all bounded on $L^p(H^n)$. In the same way the operators

$$Q_k f = \int_0^{\infty} f * e_k^{\lambda} d\mu(\lambda) - \int_0^{\infty} f * e_k^{-\lambda} d\mu(\lambda)$$

are bounded on $L^p(H^n)$. This is so because Q_k and R_k are related via the Hilbert transform H on $\mathbb{R}$. In fact, if we let

$$(Hf)^{\lambda}(z) = sign(\lambda) f^{\lambda}(z)$$

then $Q_k = R_k H$. Therefore, in order to estimate the operator norms of R_k, it is enough to do the same for Q_k. The advantage of Q_k over R_k is that it has an odd kernel, and so the method of rotations can be applied to get norm estimates.

In order to get the required norm estimates of the operators Q_k, we use the following theorem due to M. Christ [12].

Theorem 2.2.3 *Let $K(z,t)$ be an odd function of t, homogeneous of degree $(-2n-2)$ on H^n. Then the operator norm of $p.v f * K$ on $L^p(H^n), 1 < p < \infty$ is bounded by $c_p \int_{\mathbb{C}^n} |K(z,1)| \, dz$.*

Proof: The idea of the proof is to reduce to the Euclidean convolution estimate that

$$p.v \int_{-\infty}^{\infty} f(x - \sqrt{|s|}, y - s) \frac{ds}{s}$$

is bounded on $L^p(\mathbb{R}^2)$ for $1 < p < \infty$. Such singular integrals along curves have been studied by several authors; see [12] and the references therein. Writing the definition of $p.v f * K(z,t)$, making a change of variables and using the fact that K is homogeneous of degree $(-2n-2)$ and is odd in t, we get

$$p.v f * K(z,t)$$

$$= \int_{\mathbb{C}^n} \left(p.v \int_{-\infty}^{\infty} f(x - \sqrt{|s|}w, t - s - \frac{1}{2}\sqrt{|s|}Im(z.\bar{w})) \frac{ds}{s} \right) K(w, 1) \, dw.$$

Therefore, it is enough to show that the L^p operator norm of

$$p.v \int_{-\infty}^{\infty} f(x - \sqrt{|s|}w, t - s - \frac{1}{2}\sqrt{|s|}Im(z.\bar{w})) \frac{ds}{s}$$

is independent of w. Now by applying a rotation we can assume that $w = (r, 0, .., 0)$, $r \in \mathbb{R}$ and then the question is the operator bound of

$$p.v \int_{-\infty}^{\infty} f(x - \sqrt{|s|}r, y, t - s - \frac{1}{2}\sqrt{|s|}ry) \frac{ds}{s}$$

on $L^p(\mathbb{R}^3)$ or equivalently, the operator bound of

$$p.v \int_{-\infty}^{\infty} f(x - \sqrt{|s|}r, t - s - \frac{1}{2}\sqrt{|s|}ry) \frac{ds}{s}$$

on $L^p(\mathbb{R}^2)$ for all values of r and y.

To reduce matters further, we make use of the two parameter family of dilations

$$\delta(r_1, r_2)f(x, y) = f(r_1 x, r_2 y)$$

which act as multiples of isometries on $L^p(\mathbb{R}^2)$. Conjugation with these isometries transform our operators to either

$$p.v \int_{-\infty}^{\infty} f(x - \sqrt{|s|}r, t - s) \frac{ds}{s}$$

if $y = 0$ or

$$p.v \int_{-\infty}^{\infty} f(x - \sqrt{|s|}r, t - s + \sqrt{|s|}) \frac{ds}{s}$$

if $y \neq 0$. This does not change the operator norm. The second integral is further transformed into the first by a change of variables. Thus everything boils down to the estimate of

$$p.v \int_{-\infty}^{\infty} f(x - \sqrt{|s|}r, t - s)\frac{ds}{s},$$

and since this is bounded on $L^p(\mathbb{R}^2)$, the theorem is proved. ∎

Using the above theorem, we can now get estimates on the operator norms of Q_k.

Proposition 2.2.4 *Let $1 < p < \infty$. Then for every $\epsilon > 0$ the estimates*

$$\|Q_k f\|_p \leq C_p k^{2n|\frac{1}{p}-\frac{1}{2}|+\epsilon}\|f\|_p$$

hold for all $f \in L^p(H^n)$.

Proof: The kernel of the operator Q_k is given by

$$K(z,t) = (2\pi)^{-n-1}|z|^{-2n-2}\left(F_k(t|z|^{-2}) - F_k(-t|z|^{-2})\right)$$

which is

$$2i(2\pi)^{-n-1}|z|^{-2n-2}Im\left(F_k(t|z|^{-2})\right).$$

Now since $\frac{1+4it}{1-4it}$ has absolute value one, it is only the constant $\frac{(n+k)!}{k!}$ that contributes any growth in k in the estimate of $\int_{\mathbb{C}^n}|K(z,1)|\,dz$. Thus the estimate from Christ's theorem is $C_p k^n$. However when $p = 2$, we know that the operator norm is one, and hence by interpolating between $p = 2$ and p close to 1 or ∞, we obtain the estimates of the proposition. ∎

Finally, we are ready to state and prove the Abel summability result for the spectral decomposition we have obtained for L^p functions. The following theorem is due to Strichartz [71].

Theorem 2.2.5 *For any $f \in L^p(H^n), 1 < p < \infty$,*

$$\lim_{r \to 1} \sum_{k=0}^{\infty} r^k \int_{-\infty}^{\infty} f * e_k^\lambda \, d\mu(\lambda) = f$$

where the limit exists in the L^p norm.

Proof: To prove the theorem it is enough to show that the operators

$$A_r f = \sum_{k=0}^{\infty} r^k \int_{-\infty}^{\infty} f * e_k^{\lambda} \, d\mu(\lambda)$$

are uniformly bounded on $L^p(H^n)$ for $0 < r < 1$ and on a dense subspace of $L^p(H^n)$ the above series converges in the norm. As we have remarked earlier, the uniform boundedness will follow if we show that the operators

$$\tilde{A}_r f = \sum_{k=0}^{\infty} r^k Q_k f$$

are uniformly bounded.

The convolution kernel of this last operator is given by

$$\sum_{k=0}^{\infty} r^k \int_0^{\infty} \left(e^{-i\lambda t} - e^{i\lambda t} \right) \varphi_k^{\lambda}(z) \, d\mu(\lambda).$$

This kernel is of the form

$$(2\pi)^{-n-1} |z|^{-2n-2} \left(F_r(t|z|^{-2}) - F_r(-t|z|^{-2}) \right)$$

where the function $F_r(t)$ is defined by

$$F_r(t) = \int_0^{\infty} e^{-i\lambda t} \left(\sum_{k=0}^{\infty} r^k L_k^{n-1}(\tfrac{\lambda}{2}) e^{-\frac{\lambda}{4}} \right) \lambda^n \, d\lambda.$$

Using the generating function identity for the Laguerre functions, we calculate that

$$F_r(t) = 4^{n+1} \Gamma(n+1)(1-r)\left(1 + r - 4(1-r)it\right)^{-n-1}.$$

By the theorem of Christ, the operator norm of $\tilde{A}_r$ is bounded by constant times

$$\int_{\mathbb{C}^n} |z|^{-2n-2} \left(|F_r(|z|^2)| + |F_r(-|z|^2)| \right) dz$$

$$\leq C(1-r) \int_{-\infty}^{\infty} |F_r(t)| \, dt.$$

The last integral is bounded by

$$C(1+r)^{-n} \int_{-\infty}^{\infty} (1 + 16t^2)^{-\frac{n+1}{2}} \, dt.$$

Hence the operator norms of $\tilde{A}_r$ are bounded by $C(1+r)^{-n}$ which proves our claim about the uniform boundedness.

By the spectral theory the Abel means $A_r f$ converge to f in the L^2 norm. On the other hand, when f is a Schwartz class function the Abel means converge uniformly. To see this observe that if p is a polynomial in two variables then

$$p(\mathcal{L}, T)(f * e_k^\lambda) = p((2k+n)|\lambda|, \lambda)(f * e_k^\lambda)$$

and consequently

$$p(\mathcal{L}, T)f(z) =$$

$$\lim_{r \to 1} \sum_{k=0}^{\infty} r^k \int_{-\infty}^{\infty} p((2k+n)|\lambda|, \lambda) f * e_k^\lambda(z, t) \, d\mu(\lambda)$$

in the L^2 norm. So by the Sobolev embedding theorem we can conclude that the Abel means converge to f uniformly whenever f is in the Schwartz class which is dense in L^p. Then by interpolating the L^2 and L^∞ convergence, we obtain the L^p convergence for $p \geq 2$.

For $1 \leq p \leq 2$, we establish the theorem by a duality argument. As a dense subspace of L^p, we choose the class of functions whose Fourier transforms are supported on a finite number of rays, i.e., functions for which

$$f = \sum_{k=0}^{N} r^k R_k f$$

for some N. It is clear that the Abel means converge for such functions. To prove the density of the subspace we need to show that for any $g \in L^{p'}$, $(g, f) = 0$ for all f in the subspace implies $g = 0$. Now we know that the operators are projections on L^2, hence on all L^p. Therefore, for any $f \in L^p$ we know that $R_k f$ is in this subspace; hence

$$(R_k g, f) = (g, R_k f) = 0$$

by hypothesis. Thus $(A_r g, f) = 0$ for all $f \in L^p$ and taking limit as $r \to 1$ and using the summability result for $L^{p'}$, we get $(g, f) = 0$ which implies $g = 0$ as desired. ∎

Corollary 2.2.6 *For any $f \in L^p$, $1 < p < \infty$ the modified Abel means*

$$\sum_{k=0}^{N^2} (1 - \frac{1}{N})^k \int_{-N}^{N} f * e_k^\lambda \, d\mu(\lambda)$$

converges to f in the norm as N tends to infinity.

Proof: The corollary follows by a minor variant of the proof of the above theorem. Again it is clear that the corollary holds for $p = 2$. The only new element in the proof is to verify the uniform boundedness in L^p of the operators

$$\sum_{k=0}^{N^2}(1 - \frac{1}{N})^k \int_{-N}^{N} f * e_k^\lambda \, d\mu(\lambda).$$

But these are obtained from the uniformly bounded operators

$$\sum_{k=0}^{N^2}(1 - \frac{1}{N})^k R_k f$$

by composing with another uniformly bounded family, namely

$$(T_N f)^\lambda(z) = \chi_{(-N,N)}(\lambda) f^\lambda(z).$$

Therefore, it is enough to show that

$$\sum_{k=0}^{N^2}(1 - \frac{1}{N})^k R_k f$$

converges in the norm. But this follows from the fact that the operator norm of the tail series

$$\sum_{k=N^2}^{\infty}(1 - \frac{1}{N})^k R_k f$$

is bounded by a constant multiple of

$$\sum_{k=N^2}^{\infty}(1 - \frac{1}{N})^k (k + 1)^{2n|\frac{1}{p} - \frac{1}{2}|+\epsilon}$$

and this goes to zero as N tends to infinity.   ∎

This corollary will be used later in the study of the injectivity of the spherical means on the Heisenberg group.

2.3   Restriction theorems for the spectral projections

Since the discovery of Stein that the Fourier transform of an L^p function has a well defined restriction to the unit sphere S^{n-1} if p is close enough to 1, various new restriction theorems have been proved in several set ups. These theorems turned out to be very useful in harmonic analysis as well as in partial differential equations. In this section we formulate and prove some restriction theorems for the spectral projections associated to the sublaplacian on the Heisenberg group.

Let us recall the Stein-Tomas restriction theorem for the Fourier transform on $\mathbb{R}^n$. If $1 \leq p \leq \frac{2(n+1)}{n+3}$ and f is a Schwartz class function then we have the a priori estimate

$$\left(\int_{S^{n-1}} |\hat{f}(\omega)|^2 \, d\sigma(\omega)\right)^{\frac{1}{2}} \leq C\|f\|_p$$

which is called the restriction theorem. The crucial step in proving the above is the following. Writing the Fourier inversion formula in polar coordinates as

$$f(x) = (2\pi)^{-\frac{n}{2}} \int_0^\infty \left(\int_{S^{n-1}} e^{i\lambda x \cdot \omega} \hat{f}(\lambda\omega) \, d\sigma(\omega)\right) \lambda^{n-1} \, d\lambda.$$

The inner integral

$$Q_\lambda f(x) = \int_{S^{n-1}} e^{i\lambda x \cdot \omega} \hat{f}(\lambda\omega) \, d\sigma(\omega)$$

is an eigenfunction of the Laplacian Δ with eigenvalue $-\lambda^2$. In terms of $Q_\lambda f$ we can write the inversion formula as

$$f(x) = (2\pi)^{-\frac{n}{2}} \int_0^\infty Q_\lambda f(x) \lambda^{n-1} \, d\lambda.$$

The operators $Q_\lambda f$ turn out to be convolutions with Bessel functions

$$Q_\lambda f = f * \varphi_\lambda(x)$$

where

$$\varphi_\lambda(x) = (\lambda|x|)^{-\frac{n}{2}+1} J_{\frac{n}{2}-1}(\lambda|x|)$$

are Bessel functions of order $(\frac{n}{2} - 1)$. In order to prove the Stein-Tomas restriction theorem, it is enough to show that

$$\|Q_\lambda f\|_{p'} \leq C_\lambda \|f\|_p, \quad 1 \leq p \leq \frac{2(n+1)}{n+3}. \tag{2.3.3}$$

Therefore, it is natural to study analogues of $Q_\lambda f$ when Δ is replaced by the sublaplacian. For analogues of $Q_\lambda f$ in other settings, we refer to Strichartz [70].

In the previous section we have established the spectral decomposition

$$f(z,t) = \sum_{k=0}^{\infty} \int_{-\infty}^{\infty} f * e_k^{\lambda}(z,t)d\mu(\lambda)$$

where each $f * e_k^{\lambda}$ is an eigenfunction of the sublaplacian with eigenvalue $(2k+n)\,|\lambda|$. Making a change of variables and defining

$$\tilde{e}_k^{\lambda}(z,t) = e_k^{\frac{\lambda}{(2k+n)}}(z,t)$$

we rewrite the above decomposition as

$$f(z,t) = \sum_{k=0}^{\infty}(2k+n)^{-n-1}\int_{-\infty}^{\infty} f * \tilde{e}_k^{\lambda}(z,t)d\mu(\lambda).$$

Now we define

$$\mathcal{P}_{\lambda}f(z,t) = \sum_{k=0}^{\infty}(2k+n)^{-n-1}f * \left(\tilde{e}_k^{\lambda} + \tilde{e}_k^{-\lambda}\right)(z,t)$$

so that the decomposition can be written as

$$f(z,t) = \int_0^{\infty} \mathcal{P}_{\lambda}f(z,t)d\mu(\lambda).$$

Observe that $\mathcal{P}_{\lambda}f$ is an eigenfunction of the sublaplacian with eigenvalue λ. Thus $\mathcal{P}_{\lambda}f$ are the analogues of $Q_{\lambda}f$ for the sublaplacian and we are interested in estimates of the form (2.3.3) for $\mathcal{P}_{\lambda}f$.

The operator $\mathcal{P}_{\lambda}f$ is given by convolution with the kernel

$$G_{\lambda}(z,t) = \sum_{k=0}^{\infty}(2k+n)^{-n-1}\left(\tilde{e}_k^{\lambda}(z,t) + \tilde{e}_k^{-\lambda}(z,t)\right). \qquad (2.3.4)$$

We first show that these operators are not bounded from $L^p(H^n)$ into $L^{p'}(H^n)$ unless $p = 1$.

Proposition 2.3.1 *There exists a Schwartz class function f such that*

$$\mathcal{P}_1 f(z,t) = c_n e^{-\frac{1}{4n}|z|^2}\cos\frac{t}{n}.$$

Proof: Choose $\varphi \in C_0^{\infty}(\mathbb{R})$ such that $\varphi = 1$ on a neighbourhood of the points $\pm\frac{1}{n}$ and $\varphi = 0$ near 0. Define a function f by

$$f(z,t) = \int_{-\infty}^{\infty}\varphi(\lambda)e^{-i\lambda t}e^{-\frac{|\lambda|}{4}|z|^2}\,d\mu(\lambda).$$

Consider the Fourier transform

$$\int_{-\infty}^{\infty} \int_{\mathbb{C}^n} f(w,t) e^{i\lambda t} e^{\frac{i}{2} Im(z.\bar{w})} \, dw \, dt$$

$$= (2\pi)^{-n-1} \varphi(\lambda) |\lambda|^n \int_{\mathbb{C}^n} e^{-\frac{|\lambda|}{4}|z|^2} e^{\frac{i}{2} Im(z.\bar{w})} \, dw.$$

Since $\varphi_0(z) = e^{-\frac{1}{4}|z|^2}$ is an eigenfunction of the symplectic Fourier transform, we get

$$\int_{-\infty}^{\infty} \int_{\mathbb{C}^n} f(w,t) e^{i\lambda t} e^{\frac{i}{2} Im(z.\bar{w})} \, dw \, dt$$

$$= (2\pi)^{-n-1} \varphi(\lambda) e^{-\frac{1}{4|\lambda|}|z|^2}.$$

This shows that the Euclidean Fourier transform of f and hence f itself is a Schwartz class function.

To calculate $\mathcal{P}_1 f(z,t)$, let us consider

$$f * e_k^\lambda(z,t) = e^{-i\lambda t} f^\lambda *_\lambda \varphi_k^\lambda(z).$$

Since $f^\lambda(z) = (2\pi)^{-n-1} \varphi(\lambda) |\lambda|^n \varphi_0^\lambda(z)$, we have

$$f * e_k^\lambda(z,t) = (2\pi)^{-n-1} e^{-i\lambda t} \varphi(\lambda) |\lambda|^n \varphi_0^\lambda *_\lambda \varphi_k^\lambda(z).$$

We then get $f * e_k^\lambda = 0$ for all $k \neq 0$ in view of the orthogonality properties of the Laguerre functions, and so

$$f * e_0^\lambda(z,t) = (2\pi)^{-1} e^{-i\lambda t} \varphi(\lambda) \varphi_0^\lambda(z).$$

Consequently,

$$f * G_\lambda(z,t)$$

$$= (2\pi)^{-1} n^{-n-1} \{ \varphi(\frac{\lambda}{n}) \varphi_0^{\frac{\lambda}{n}}(z) e^{-i\frac{\lambda}{n} t} + \varphi(\frac{-\lambda}{n}) \varphi_0^{\frac{\lambda}{n}}(z) e^{i\frac{\lambda}{n} t} \}$$

and taking $\lambda = 1$ and noting that $\varphi = 1$ near the points $\pm \frac{1}{n}$ we get the proposition with $c_n = \pi^{-1} n^{-n-1}$. ∎

From the above proposition it is clear that $\mathcal{P}_\lambda$ cannot be bounded from L^p into $L^{p'}$ unless $p = 1$. So we consider the mixed norm spaces

$$L^{(p,r)}(H^n) = L^p(\mathbb{C}^n, L^r(\mathbb{R}))$$

with the norm

$$\|f\|_{(p,r)}^p = \int_{\mathbb{C}^n} \left(\int_{-\infty}^{\infty} |f(z,t)|^r \, dt \right)^{\frac{p}{r}} \, dz.$$

We can ask if $\mathcal{P}_\lambda$ is bounded from $L^{(p,r)}(H^n)$ into $L^{(p',r')}(H^n)$. Again from the proposition, we infer that this cannot be true unless $r = 1$. When $r = 1$ we have the following result due to Müller [44].

Theorem 2.3.2 *If $1 \leq p < 2$ and $f \in L^{(p,r)}(H^n)$, then we have the inequality*

$$\|\mathcal{P}_\lambda f\|_{(p',\infty)} \leq C_\lambda \|f\|_{(p,1)}.$$

Proof: The proof of this theorem is fairly simple in comparison with the proof of the Stein-Tomas restriction theorem. For the sake of simplicity of notation, let us take $f(z,t) = h(t)g(z)$. Recalling the definition of $\mathcal{P}_\lambda f$ we have

$$\mathcal{P}_\lambda f(z,t) = \sum_{k=0}^{\infty} (2k+n)^{-n-1} e^{-i\lambda_k t} \hat{h}(\lambda_k) g *_{\lambda_k} \varphi_k^{\lambda_k}(z)$$

$$+ \sum_{k=0}^{\infty} (2k+n)^{-n-1} e^{i\lambda_k t} \hat{h}(-\lambda_k) g *_{-\lambda_k} \varphi_k^{\lambda_k}(z)$$

where we have written $\lambda_k = \frac{\lambda}{(2k+n)}$. Therefore, in order to prove the theorem, it is enough to show that

$$\sum_{k=0}^{\infty} (2k+n)^{-n-1} \left(\|g *_{-\lambda_k} \varphi_k^{\lambda_k}\|_{p'} + \|g *_{\lambda_k} \varphi_k^{\lambda_k}\|_{p'} \right) \qquad (2.3.5)$$

$$\leq C\|g\|_p.$$

So we need to get estimates of each term in the above series.

Consider the twisted convolution $g *_\lambda \varphi_k^\lambda$ for $\lambda > 0$. Writing the definition and making a change of variables

$$g *_\lambda \varphi_k^\lambda \left(\frac{z}{\sqrt{\lambda}} \right)$$

$$= \lambda^{-n} \int_{\mathbb{C}^n} g \left(\frac{z-w}{\sqrt{\lambda}} \right) \varphi_k(w) e^{\frac{i}{2} Im(z.\bar{w})} \, dw.$$

So, with $g_\lambda(z) = g(\frac{z}{\sqrt{\lambda}})$ we have

$$g *_\lambda \varphi_k^\lambda \left(\frac{z}{\sqrt{\lambda}} \right) = \lambda^{-n} g_\lambda \times \varphi_k(z).$$

From this we get

$$\|g *_\lambda \varphi_k^\lambda\|_{p'} \leq C\lambda^{-n-\frac{n}{p'}} \|g_\lambda \times \varphi_k\|_{p'}$$

and so it is necessary to get estimates on the projections $f \to f \times \varphi_k$.

Since $g \times \varphi_k$ is the projection of g onto the kth eigenspace of the special Hermite operator, we immediately get

$$\|g \times \varphi_k\|_2 \le \|g\|_2.$$

Since $\|\varphi_k\|_\infty \le Ck^{n-1}$ we also have

$$\|g \times \varphi_k\|_\infty \le Ck^{n-1}\|g\|_1.$$

Interpolation between these two estimates gives us

$$\|g \times \varphi_k\|_{p'} \le Ck^{(n-1)(1-\frac{2}{p'})}\|g\|_p.$$

In view of this estimate we have

$$\|g *_\lambda \varphi_k^\lambda\|_{p'} \le C\lambda^{-\frac{2n}{p'}}k^{(n-1)(1-\frac{2}{p'})}\|g\|_p.$$

Using these estimates for $\lambda = \lambda_k$ in (2.3.5) we see that the series is dominated by

$$C_\lambda\|g\|_p \sum_{k=0}^{\infty}(2k+n)^{-n-1}(2k+n)^{\frac{2n}{p'}+(n-1)(1-\frac{2}{p'})}.$$

The last series reduces to $\sum_{k=0}^{\infty}(2k+n)^{-2+\frac{2}{p'}}$ which converges whenever $p < 2$. This completes the proof of the restriction theorem. ∎

We now proceed to prove an improvement of the above restriction theorem. In the case of the Fourier transform on $\mathbb{R}^2$, Zygmund [93] has obtained the following improvement of the Stein-Tomas restriction theorem. If $f \in L^p(\mathbb{R}^2)$, $1 \le p \le \frac{4}{3}$,$q = \frac{1}{3}p'$ then for any $\lambda > 0$,

$$\left(\int_{|x|=\lambda}|\hat{f}(x)|^q \, d\sigma(x)\right)^{\frac{1}{q}} \le C\lambda^{\frac{1}{p'}}\|f\|_p.$$

Similar results are known in higher dimensions as well. (See the results in Chapter IX of Stein [65].) We would like to prove the improvement of Müller's restriction theorem. Here is one such result.

Theorem 2.3.3 *Let $0 < \gamma < \frac{3n-2}{3n+4}$ and $1 \le p \le 1 + \gamma$. Then with $q = \frac{2\gamma}{1+\gamma}p'$ we have*

$$\|P_\lambda f\|_{(q,\infty)} \le C_\lambda\|f\|_{(p,1)}$$

for all $f \in L^{(p,1)}(H^n)$. When we consider only radial functions, γ can be taken in the range $0 < \gamma < \frac{2n-1}{2n+1}$.

In the proof of Müller's theorem we used the estimate

$$\|f \times \varphi_k\|_{p'} \leq Ck^{(n-1)(1-\frac{2}{p'})}\|f\|_p$$

which was obtained by interpolating the trivial end point estimates. In order to prove the above version of the restriction theorem we need sharp estimates of the projections $f \to f \times \varphi_k$ which are given in the following proposition.

Proposition 2.3.4 *(i) For* $1 \leq p < \frac{2(3n+1)}{3n+4}$ *we have*

$$\|f \times \varphi_k\|_{p'} \leq Ck^{2n(\frac{1}{p}-\frac{1}{2})-1}\|f\|_p.$$

(ii) If we consider only radial functions then the above estimates are valid in the larger range $1 \leq p < \frac{4n}{2n+1}$.

We will assume the proposition for a moment and prove the theorem first. As

$$f \to (2\pi)^{-n}(f \times \varphi_k)$$

are spectral projections we have

$$\|f \times \varphi_k\|_2^2 = (f \times \varphi_k, f \times \varphi_k) = (2\pi)^n(f \times \varphi_k, f)$$

and applying Hölder's inequality and using the estimates of the proposition we get

$$\|f \times \varphi_k\|_2^2 \leq Ck^{2n(\frac{1}{p}-\frac{1}{2})-1}\|f\|_p^2.$$

Thus we have

$$\|f \times \varphi_k\|_2 \leq Ck^{n(\frac{1}{p}-\frac{1}{2})-\frac{1}{2}}\|f\|_p$$

which holds for $1 \leq p \leq (1+\gamma)$, $0 < \gamma < \frac{3n-2}{3n+4}$. Notice that when

$$p = p_\gamma = (1+\gamma), \quad \frac{2\gamma}{1+\gamma}p' = 2$$

so that we have

$$\|f \times \varphi_k\|_{\frac{2\gamma}{1+\gamma}p'_\gamma} \leq Ck^{n(\frac{1}{p}-\frac{1}{2})-\frac{1}{2}}\|f\|_{p_\gamma}.$$

We also have

$$\|f \times \varphi_k\|_\infty \leq Ck^{n-1}\|f\|_1.$$

Interpolating these two inequalities we obtain

$$\|f \times \varphi_k\|_q \leq Ck^{(n-1)(1-\frac{1}{q})-\frac{2n}{q}\frac{\gamma}{1+\gamma}}\|f\|_p$$

for $1 \leq p \leq (1 + \gamma)$ where $q = \frac{2\gamma}{1+\gamma} p'$.

Proceeding as in the proof of Theorem 2.3.2, we get for $f(z,t) = h(t)g(z)$ the estimate

$$\|\mathcal{P}_\lambda f\|_{(q,\infty)} \leq C\|h\|_1$$

$$\times \sum_{k=0}^{\infty} (2k + n)^{-n-1} \left(\|g *_{\lambda_k} \varphi_k^{\lambda_k}\|_q + \|g *_{-\lambda_k} \varphi_k^{\lambda_k}\|_q \right).$$

Using the above estimates we get

$$\|g *_{\lambda_k} \varphi_k^{\lambda_k}\|_q \leq C\lambda_k^{-\frac{n}{p'}-\frac{n}{q}} k^{(n-1)(1-\frac{1}{q})-\frac{2n}{q}\frac{\gamma}{1+\gamma}} \|g\|_p.$$

Therefore, the theorem will be established if we show that the series

$$\sum_{k=0}^{\infty} (2k + n)^{-n-1+\frac{n}{p'}+\frac{n}{q}} (2k + n)^{(n-1)(1-\frac{1}{q})-\frac{2n}{q}\frac{\gamma}{1+\gamma}}$$

converges. A calculation shows that the series converges precisely when $q > 1$ which is guaranteed whenever $1 \leq p \leq (1+\gamma)$. For radial functions we use the estimates (ii) of the proposition. $\blacksquare$

Coming to the proof of the proposition we already know that the result is true when $p = 1$. We will show that the estimates are valid when $p = (1 + \gamma)$ for $0 < \gamma < \frac{3n-2}{3n+4}$. Then by Riesz-Thorin we will get the required estimates for $1 \leq p \leq (1 + \gamma)$. In order to prove the result at $p = (1 + \gamma)$, we use analytic interpolation. Let

$$\psi_k^\alpha(z) = \frac{\Gamma(k+1)\Gamma(\alpha+1)}{\Gamma(k+\alpha+1)} L_k^\alpha(\frac{1}{2}|z|^2)e^{-\frac{1}{4}|z|^2}$$

be the Laguerre functions of type α. These functions can be defined for even complex values of α, $Re(\alpha) > -1$, and using them we define the following analytic family of operators. We set

$$G_k^\alpha f(z) = f \times \psi_k^{-\frac{1}{3}+(n+\frac{1}{3})\alpha}(z).$$

Note that when $\alpha = \frac{3n-2}{3n+1}$ we recover $f \times \varphi_k$:

$$G_k^{\frac{3n-2}{3n+1}} f(z) = \frac{\Gamma(k+1)\Gamma(n)}{\Gamma(k+n)} f \times \varphi_k(z).$$

For this analytic family we will establish

Proposition 2.3.5 *Let $\alpha = \sigma + i\tau, 0 < \sigma \leq n$. Then we have*

$$(i) \qquad \|G_k^{\sigma+i\tau} f\|_\infty \leq C(1 + |\tau|)^{\frac{2}{3}} \|f\|_1,$$

$$(ii) \qquad \|G_k^{1+i\tau} f\|_2 \leq C(1 + |\tau|)^n k^{-n} \|f\|_2.$$

Once we have this proposition, the main estimate, namely,

$$\|f \times \varphi_k\|_{p'} \leq C k^{2n(\frac{1}{p} - \frac{1}{2}) - 1} \|f\|_p$$

with $p = (1 + \gamma)$ follows by analytic interpolation. The end point estimates for $Re(\alpha) > 0$ and $Re(\alpha) = 1$ are provided by the proposition.

In order to prove the end point estimates, we use several properties of the Laguerre functions. First of all, Laguerre polynomials satisfy the relation

$$L_k^{\alpha+\beta}(t) = \sum_{j=0}^{k} A_{k-j}^{\beta} L_{k-j}^{\alpha}(t)$$

where A_k^β are the binomial coefficients defined by

$$A_k^\beta = \frac{\Gamma(k + \beta + 1)}{\Gamma(k + 1)\Gamma(\beta + 1)}.$$

This formula can be proved by using the generating function identity satisfied by the Laguerre functions; see Szego [72] or [84]. In view of the above relation we get

$$\psi_k^{n+i\tau}(z) = \frac{\Gamma(k + 1)\Gamma(n + 1 + i\tau)}{\Gamma(k + n + 1 + i\tau)} \sum_{j=0}^{k} A_{k-j}^{i\tau} \varphi_j(z).$$

Since $W(\varphi_j) = (2\pi)^n P_j$, we get the formula

$$W(\psi_k^{n+i\tau}) = (2\pi)^n \frac{\Gamma(k + 1)\Gamma(n + 1 + i\tau)}{\Gamma(k + n + 1 + i\tau)} \sum_{j=0}^{k} A_{k-j}^{i\tau} P_j.$$

Appealing to the Plancherel theorem for the Weyl transform and noting that

$$\|TS\|_{HS} \leq \|T\| \|S\|_{HS}$$

where $\|T\|$ is the operator norm, we see that

$$\|G_k^{1+i\tau} f\|_2 \leq C \left| \frac{\Gamma(k + 1)\Gamma(n + 1 + i(n + \frac{1}{3})\tau)}{\Gamma(k + n + 1 + i\tau)} \right|$$

$$\times \| \sum_{j=0}^{k} A_{k-j}^{i\tau} P_j \| \| f \|_2.$$

As P_j are projections

$$\| \sum_{j=0}^{k} A_{k-j}^{i(n+\frac{1}{3})\tau} P_j \| \leq \max_{1 \leq j \leq k} \{ |A_j^{i(n+\frac{1}{3})\tau}| \}$$

which gives the estimate

$$\| G_k^{1+i\tau} f \|_2 \leq C \max_{1 \leq j \leq k} \{ |A_j^{i(n+\frac{1}{3})\tau}| |A_k^{n+i(n+\frac{1}{3})\tau}|^{-1} \} \| f \|_2.$$

Using Stirling's formula for the Gamma function it is not difficult to show that

$$|A_j^{i(n+\frac{1}{3})\tau}| |A_k^{n+i(n+\frac{1}{3})\tau}|^{-1} \leq C(1 + |\tau|)^{(n+\frac{1}{3})} k^{-n}$$

for all $1 \leq j \leq k$. This completes the proof of estimate (ii) of Proposition 2.3.5.

Finally, the estimate (i) of the proposition follows from the next lemma.

Lemma 2.3.6 *Let $\alpha = \sigma + i\tau, -1 < \sigma \leq n$. Then we have*

$$|\psi_k^{\alpha}(z)| \leq C(1 + |\tau|)^{\frac{2}{3}}$$

for all $\sigma > -\frac{1}{3}$ with a constant C independent of z and k.

Proof: To get the estimate of the lemma we use the following formula which connects Laguerre polynomials of different type (see Erdelyi et al [20]):

$$L_k^{\alpha+\beta}(t) = \frac{\Gamma(k + \alpha + \beta + 1)}{\Gamma(\beta)\Gamma(k + \alpha + 1)} \int_0^1 s^{\alpha}(1 - s)^{\beta-1} L_k^{\alpha}(ts) \, ds.$$

From the above formula we get

$$\psi_k^{\alpha}(z) = \frac{\Gamma(\alpha + 1)}{\Gamma(-\frac{1}{3})\Gamma(\alpha + \frac{1}{3})}$$

$$\times \left(\int_0^1 s^{-\frac{1}{3}}(1 - s)^{\alpha-\frac{2}{3}} \psi_k^{-\frac{1}{3}}(\sqrt{s}z) e^{-\frac{1}{4}(1-s)|z|^2} \, ds \right)$$

which is valid for $Re(\alpha) > -\frac{1}{3}$. Now, as can be seen from the asymptotic properties of the Laguerre functions given in Chapter VIII of Szego [72], the Laguerre functions $\psi_k^\alpha(z)$ are uniformly bounded as long as $\alpha \geq -\frac{1}{3}$ and therefore,

$$\sup_z |\psi_k^\alpha(z)| \leq C \frac{|\Gamma(\alpha + i\tau + 1)|}{|\Gamma(\alpha + i\tau + \frac{1}{3})|}$$

for all $\sigma > -\frac{1}{3}$. Gamma function estimates show that

$$\frac{|\Gamma(\alpha + i\tau + 1)|}{|\Gamma(\alpha + i\tau + \frac{1}{3})|} \leq C(1 + |\tau|)^{\frac{2}{3}}$$

and this completes the proof of the lemma.   ∎

Thus we have established the estimates (i) of the Proposition 2.3.4. To prove part (ii) of that proposition we use the fact that when f is radial

$$f \times \varphi_k = R_k(f)\varphi_k.$$

Therefore, by Hölder's inequality

$$\|f \times \varphi_k\|_{p'} \leq Ck^{-(n-1)}\|\varphi_k\|_{p'} \cdot \|\varphi_k\|_p \|f\|_p.$$

Sharp estimates for the L^p norms of the Laguerre functions are known:

$$\|\varphi_k\|_p \leq Ck^{\frac{n}{p}-\frac{1}{2}}$$

when $p < \frac{4n}{2n-1}$ and when $p > \frac{4n}{2n-1}$,

$$\|\varphi_k\|_p \leq Ck^{n-1-\frac{n}{p}}.$$

Using these estimates we easily get part (ii) of the Proposition 2.3.4. For a proof of these estimates we refer to [84]; see Lemma 1.5.4.   ∎

Consider now the spectral decomposition $\mathcal{L} = \int_0^\infty \lambda \, dE(\lambda)$ of the sublaplacian. Let

$$E_{a,b}f = E(b)f - E(a)f, \quad 0 \leq a \leq b$$

be the spectral projections associated to the interval $a \leq \lambda \leq b$. We encounter these projections in the study of Bochner-Riesz means for the sublaplacian. In terms of $\mathcal{P}_\lambda$ we can write

$$E_{a,b}f = \int_a^b \mathcal{P}_\lambda f \, d\mu(\lambda)$$

and we are interested in $L^p - L^2$ mapping properties of these projections. We have

Theorem 2.3.7 *For $1 \le p \le 2$ we have the inequalities*

$$(i) \qquad \|E_{a,b}f\|_2 \le C_p((b-a)b^n)^{(\frac{1}{p}-\frac{1}{2})}\|f\|_p,$$

$$(ii) \qquad \|E_{a,b}f\|_{p'} \le C_p((b-a)b^n)^{2(\frac{1}{p}-\frac{1}{2})}\|f\|_p$$

for all $f \in L^p(H^n)$. Moreover, when $b < 2a$ the estimates are sharp.

Proof: One can easily see that (i) and (ii) are equivalent. $E_{a,b}$ are given by the kernel

$$G_{a,b}(z,t) = (2\pi)^{-n-1} \sum_{k=0}^{\infty} \int_{\frac{a}{2k+n}}^{\frac{b}{2k+n}} (e^{-i\lambda t} + e^{i\lambda t})\varphi_k^\lambda(z)\lambda^n \, d\lambda.$$

Applying the Plancherel theorem in the t variable, using the orthogonality of φ_k^λ, and noting that

$$\|\varphi_k^\lambda\|_2 \le C\lambda^{-\frac{n}{2}}k^{\frac{n-1}{2}},$$

we get the estimate

$$\|G_{a,b}\|_2^2 \le C \int_a^b \lambda^n \, d\lambda$$

$$\le C(b^{n+1} - a^{n+1}) \le C((b-a)b^n).$$

This implies that

$$\|E_{a,b}f\|_2 \le C((b-a)b^n)^{\frac{1}{2}}\|f\|_1.$$

Interpolation with the estimate $\|E_{a,b}f\|_2 \le \|f\|_2$ proves (i).

To show that these estimates are sharp when $b < 2a$, choose $\varphi \in C_0^\infty(\mathbb{R})$ such that $0 \le \varphi \le 1$, $\varphi = 1$ on a neighbourhood of the interval $I = [\frac{a}{n}, \frac{b}{n}]$ and $\varphi = 0$ outside a sufficiently small open interval containing I and such that φ' changes sign only once. Define

$$f(z,t) = \int_0^\infty \varphi(\lambda)e^{-i\lambda t}e^{-\frac{\lambda}{4}|z|^2} \, d\mu(\lambda).$$

Integrating by parts we also have

$$f(z,t) = -\frac{i}{t}(2\pi)^{-n-1} \int_0^\infty e^{-i\lambda t} \frac{d}{d\lambda}\left(\varphi(\lambda)e^{-\frac{\lambda}{4}|z|^2}\lambda^n\right) \, d\lambda.$$

The first formula for f implies that

$$\int_{\mathbb{C}^n} \int_{|t|(b-a)<1} |f(z,t)|^p \, dz \, dt$$

$$\le C(b-a)^{-1}(b^{n+1} - a^{n+1})^p a^{-n}.$$

Similarly the second formula for f implies

$$\int_{\mathbb{C}^n} \int_{|t|(b-a)\ge 1} |f(z,t)|^p \, dz \, dt$$

$$\le C(b-a)^{p-1} \left(b^{n(p-1)} + (b^n - a^n)^p a^{-n} \right)$$

whenever $p > 1$. If we assume $b < 2a$ then the last two estimates yield

$$\|f\|_p \le C((b-a)b^n)^{p-1}.$$

On the other hand, we have

$$\|E_{a,b}f\|_2^2 = (2\pi)^{-n-1} \int_{\frac{a}{n}}^{\frac{b}{n}} \lambda^n \, d\lambda \ge C(b^{n+1} - a^{n+1}).$$

Therefore, we have

$$\|E_{a,b}f\|_2 \ge C((b-a)b^n)^{\frac{1}{p}-\frac{1}{2}} \|f\|_p.$$

If $p = 1$ the same estimate can be obtained by another integration by parts. The above shows that the estimates are sharp.   ∎

We remark that the simple method of estimating $E_{a,b}f$ used above, which only makes use of the size of $G_{a,b}(z,t)$ would not yield the best possible result in the case of the Laplacian on $\mathbb{R}^n$.

Recall that $\mathcal{P}_\lambda f = f * G_\lambda$ where G_λ is explicitly given by (2.3.4). If δ_r are the nonisotropic dilations, then it is easily seen that

$$(f * g)_r = f_r * g_r$$

where

$$f_r(z,t) = r^{-Q} f(\delta_{r^{-1}}(z,t)),$$

with $Q = (2n + 2)$ being the homogeneous dimension of the Heisenberg group. The explicit formula for G_λ shows that

$$G_\lambda = \lambda^{-1}(G_1)_{\lambda^{-\frac{1}{2}}}$$

and consequently, we get

$$\|\mathcal{P}_\lambda f\|_{(p',\infty)} \le C_p \lambda^{n(\frac{2}{p}-1)} \|f\|_{(p,1)}$$

for $1 \le p < 2$. In view of this we also have the estimate

$$\|E_{a,b}f\|_{(p',\infty)} \le C(b^{n(\frac{2}{p}-1)+1} - a^{n(\frac{2}{p}-1)+1}) \|f\|_{(p,1)}.$$

Therefore, by interpolation we obtain

Corollary 2.3.8 *If* $1 \le p \le 2$ *and* $1 \le r \le p$ *then*

$$\|E_{a,b}f\|_{(p',r')} \le C_p(b-a)^{(\frac{2}{r}-1)}b^{n(\frac{2}{p}-1)}\|f\|_{(p,r)}$$

for all $f \in L^{(p,r)}(H^n)$.

In the above restriction theorem we considered integrals of $\mathcal{P}_\lambda f$ over finite intervals. Earlier we considered integrals of the individual projections $f * e_k^\lambda$ over the entire real line. We now consider integrals of $f * e_k^\lambda$ over finite intervals and for such projections we prove a restriction theorem. We start with a simple lemma.

Lemma 2.3.9 *For* $f \in L^p(\mathbb{R}), 1 \le p \le 2$, *we have*

$$\left(\int_{-a}^{a} |\hat{f}(\lambda)|^2 \, d\lambda\right)^{\frac{1}{2}} \le Ca^{(\frac{1}{p}-\frac{1}{2})}\|f\|_p.$$

Proof: As the lemma is clearly true for $p = 2$, we assume $p < 2$. Let χ_a be the characteristic function of the interval $-a \le t \le a$ so that

$$\hat{\chi}_a(\lambda) = (\frac{2}{\pi})^{\frac{1}{2}}\frac{\sin(a\lambda)}{\lambda}.$$

By Plancherel and Young inequalities

$$\left(\int_{-a}^{a} |\hat{f}(\lambda)|^2 \, d\lambda\right)^{\frac{1}{2}} = \left(\int_{-\infty}^{\infty} |f * \hat{\chi}_a(t)|^2 \, dt\right)^{\frac{1}{2}}$$

which is dominated by

$$\|f\|_p\|\hat{\chi}_a\|_q, \qquad \frac{1}{p} + \frac{1}{q} - 1 = \frac{1}{2}.$$

But

$$\|\hat{\chi}_a\|_q = (\frac{2}{\pi})^{\frac{1}{2}}a \left(\int_{-\infty}^{\infty} |\frac{\sin(a\lambda)}{a\lambda}|^q \, d\lambda\right)^{\frac{1}{q}}$$

which equals a constant times $a^{1-\frac{1}{q}} = a^{(\frac{1}{p}-\frac{1}{2})}$ and the lemma follows. ∎

For $a > 0$, we now define the operators

$$P_{k,a}f = \int_{-a}^{a} f * e_k^\lambda \, d\mu(\lambda).$$

For these operators we have

Theorem 2.3.10 *Let* $1 \leq p < \frac{2(3n+1)}{3n+4}$ *and* $f \in L^p(H^n)$. *Then*

$$\|P_{k,a}f\|_2 \leq Ck^{n(\frac{1}{p}-\frac{1}{2})-\frac{1}{2}}a^{(n+1)(\frac{1}{p}-\frac{1}{2})}\|f\|_p.$$

Proof: As

$$f * e_k^\lambda(z,t) = e^{-i\lambda t}f^\lambda *_\lambda \varphi_k^\lambda(z),$$

by applying the Plancherel theorem in the t variable we have

$$\|P_{k,a}f\|_2^2 = C\int_{\mathbb{C}^n}\int_{-a}^a |f^\lambda *_\lambda \varphi_k^\lambda(z)|^2 \lambda^{2n}\, d\lambda\, dz.$$

Using the estimates of Proposition 2.3.4 we get

$$\int_{\mathbb{C}^n} |f^\lambda *_\lambda \varphi_k^\lambda(z)|^2\, dz \leq Ck^{2n(\frac{1}{p}-\frac{1}{2})-1}\lambda^{-3n+\frac{2n}{p}}\|f^\lambda\|_p^2.$$

Thus we have

$$\|P_{k,a}f\|_2^2 =$$

$$Ck^{2n(\frac{1}{p}-\frac{1}{2})-1}\int_{-a}^a |\lambda|^{-n+\frac{2n}{p}}\left(\int_{\mathbb{C}^n} |f^\lambda(z)|^p\, dz\right)^{\frac{2}{p}} d\lambda.$$

Applying Minkowski's integral inequality we get

$$\left(\int_{-a}^a |\lambda|^{-n+\frac{2n}{p}}\left(\int_{\mathbb{C}^n}|f^\lambda(z)|^p\, dz\right)^{\frac{2}{p}} d\lambda\right)^{\frac{p}{2}}$$

$$\leq C\int_{\mathbb{C}^n} dz\left(\int_{-a}^a |f^\lambda(z)|^2|\lambda|^{-n+\frac{2n}{p}}\, d\lambda\right)^{\frac{p}{2}}.$$

Using the lemma the last integral is bounded by

$$a^{n-\frac{np}{2}}\int_{\mathbb{C}^n}\left(\int_{-a}^a |f^\lambda(z)|^2\, d\lambda\right)^{\frac{p}{2}}$$

$$\leq Ca^{n-\frac{np}{2}}a^{1-\frac{p}{2}}\left(\int_{H^n} |f(z,t)|^p\, dz\, dt\right).$$

Finally, we get the estimate

$$\|P_{k,a}f\|_2^2 \leq Ck^{2n(\frac{1}{p}-\frac{1}{2})-1}a^{2(n+1)(\frac{1}{p}-\frac{1}{2})}\|f\|_p^2$$

and this proves the theorem.   ∎

2.4 A Paley-Wiener theorem for the spectral projections

Besides the usual Paley-Wiener theorem for the Fourier transform on $\mathbb{R}^n$, there is another 'Paley-Wiener type' theorem for the spectral projections $f * \varphi_\lambda(x)$ studied by Strichartz. In a long paper [70], Strichartz studied on various spaces (e.g. hyperbolic spaces, semi Riemannian spaces) how the properties of the function f relate to the properties of the spectral projections $P_\lambda f$ associated to the underlying Laplacian. Among other things he obtained 'Paley-Wiener theorems' for the operators $P_\lambda f$ in the case of Euclidean and semi Riemannian symmetric spaces. In a subsequent paper [71], in which Strichartz developed the L^p spectral theory of the sublaplacian on the Heisenberg group, he raises the question of proving a Paley-Wiener theorem for the projections $f \to f * e_k^\lambda$ on the Heisenberg group. We prove one such theorem in this section.

In order to state our result, we need to introduce some notation. Let Δ_+ and Δ_- be the usual forward and backward finite difference operators, namely,

$$\Delta_+ \psi(k) = \psi(k+1) - \psi(k),$$

and

$$\Delta_- \psi(k) = \psi(k) - \psi(k-1).$$

We then define a second order finite difference operator, denoted by Δ, by

$$\Delta\psi(k) = k\Delta_+\Delta_-\psi(k) + n\Delta_+\psi(k).$$

This operator occurs naturally when we try to relate the Laguerre coefficients of the functions $r^2 \, f(r)$ and $f(r)$. In fact, let $f(r)$ have the Laguerre expansion

$$f(r) = \sum_{k=0}^{\infty} \psi(k)\varphi_k(r)$$

where

$$\varphi_k(r) = L_k^{n-1}(\frac{1}{2}r^2)e^{-\frac{1}{4}r^2}$$

and consider the effect of multiplying the above series by $\frac{1}{2}r^2$. Using the recursion formula (see 5.1.10 Szego [72])

$$rL_k^{n-1}(r) =$$

$$(2k + n)L_k^{n-1}(r) - (k + 1)L_{k+1}^{n-1}(r) - (k + n - 1)L_{k-1}^{n-1}(r)$$

and rearranging the series we get

$$\frac{1}{2}r^2 f(r) = \sum_{k=0}^{\infty} \Delta \psi(k) \varphi_k(r).$$

This second order finite difference operator plays an important role in the formulation and the proof of our Paley-Wiener theorem.

For each $j \geq 1$, we define Δ^j inductively. Using these operators we define the sequence spaces l_j^2. We say that a sequence $\psi = (\psi(k))$ belongs to l_j^2 if the semi norm

$$\|\psi\|_{2,j} = \left(\sum_{k=0}^{\infty} |\Delta^j \psi(k)|^2 \right)^{\frac{1}{2}}$$

is finite. In what follows we assume that $n = 1$ for the sake of simplicity of notation. The main theorem of this section can be formulated and proved even when $n \geq 2$, but we leave this to the interested reader.

Let f be a continuous function on H^1. For an integer m, we let

$$f_m(z, t) = z^m f(z, t)$$

when $m \geq 0$ and when $m \leq 0$, we let

$$f_m(z, t) = \bar{z}^m f(z, t).$$

We also let

$$\psi_m(k, z, t, \lambda) = f_m * e_k^\lambda(z, t).$$

Theorem 2.4.1 *The function f on H^1 is supported in $|z| \leq B$ if and only if for every m and $j \geq 0$ the sequence $\psi_m = (\psi_m(k, z, t, \lambda)) \in l_j^2$ and*

$$\|\psi_m\|_{2,j} \leq C_m |\lambda|^{j-1} 2^{-j} (B + |z|)^{2j}.$$

As we observe, the above is essentially a theorem for the $z-$ variable and therefore, we may (and we will) assume that $\lambda = 1$. Then we are interested in characterising compactly supported functions f on $\mathbb{C}^n$ in terms of properties of the spectral projections $f \to f \times \varphi_k$. We start with a simple proposition. Recall the definitions of the derivations δ_j and $\bar{\delta}_j$ introduced in Section 1.5. Define

$$D = \sum_{j=1}^{n} (\delta_j \bar{\delta}_j + \bar{\delta}_j \delta_j)$$

which acts on bounded operators defined on $L^2(\mathbb{R}^n)$.

Proposition 2.4.2 *A function $f \in L^2(\mathbb{C}^n)$ is supported in $|z| \le B$ if and only if for each j, $D^j W(f)$ is a Hilbert-Schmidt operator satisfying the estimate*

$$\|D^j W(f)\|_{HS} \le CB^{2j}$$

where C is a fixed constant.

The proposition is an immediate consequence of the relation

$$D^j W(f) = W(|z|^{2j} f)$$

which follows from the formulas

$$\delta_j W(f) = W(\bar{z}_j f), \quad \bar{\delta}_j W(f) = W(z_j f)$$

and the Plancherel theorem for the Weyl transform. We now specialise to the case of radial functions.

When f is radial we know that the Weyl transform reduces to a Laguerre transform and $W(f) = m(H)$ where

$$m(2k + n) = \frac{k!(n-1)!}{(k+n-1)!} \int_{\mathbb{C}^n} f(z) \varphi_k(z) \, dz.$$

So the special Hermite expansion of f reduces to the Laguerre series

$$f(z) = (2\pi)^{-n} \sum_{k=0}^{\infty} m(2k+n) \varphi_k(z)$$

where the series converges in the L^2 norm. We let $\psi(k) = m(2k+n)$ and calculate the effect of multiplying f by $\frac{1}{2}|z|^2$ on $\psi(k)$. As we have seen, we get the relation

$$\frac{1}{2}|z|^2 f(z) = -(2\pi)^{-n} \sum_{k=0}^{\infty} \Delta\psi(k)\varphi_k(z).$$

An iteration produces the formula

$$2^{-j}|z|^{2j} f(z) = (-1)^j (2\pi)^{-n} \sum_{k=0}^{\infty} \Delta^j \psi(k)\varphi_k(z). \qquad (\,2.4.6\,)$$

We now have the following theorem for radial functions.

Theorem 2.4.3 *For any radial function $f \in L^2(\mathbb{C})$, the following conditions are equivalent: (i) f is supported in $|z| \le B$ (ii) $(\psi(k)) \in l_j^2$ and $\|\psi\|_{2,j} \le C2^{-j}B^{2j}$ for all $j \ge 0$ where $\psi(k)$ are the Laguerre coefficients of f.*

Proof: In view of the formula (2.4.6) and the orthogonality of the functions φ_k, we get

$$2^{-2j} \int_{\mathbb{C}} |z|^{4j} |f(z)|^2 \, dz = (2\pi)^{-2} \sum_{k=0}^{\infty} |\Delta^j \psi(k)|^2 \int_{\mathbb{C}} (\varphi_k(z))^2 \, dz.$$

But $\int_{\mathbb{C}} (\varphi_k(z))^2 \, dz$ is a constant independent of k. Therefore,

$$2^{-2j} \int_{\mathbb{C}} |z|^{4j} |f(z)|^2 \, dz = C(2\pi)^{-2} \sum_{k=0}^{\infty} |\Delta^j \psi(k)|^2$$

and so, the right hand side is less than or equal to $2^{-2j} B^{4j}$ if and only if

$$\int_{\mathbb{C}} |z|^{4j} |f(z)|^2 \, dz \leq C B^{4j}.$$

This is the case for all $j \geq 0$ when f is supported in $|z| \leq B$.   ∎

In order to prove our main theorem we need one more result. Let μ_r stand for the normalised surface measure on the sphere

$$S_r = \{ z \in \mathbb{C}^n : |z| = r \}.$$

The following theorem gives us the expansion of $f \times \mu_r$ in terms of $f \times \varphi_k$.

Theorem 2.4.4 *For $f \in L^2(\mathbb{C}^n)$, we have the expansion*

$$f \times \mu_r(z) = (2\pi)^{-n} \sum_{k=0}^{\infty} \frac{k!(n-1)!}{(k+n-1)!} \varphi_k(r) f \times \varphi_k(z)$$

where $\varphi_k(r)$ stands for $L_k^{n-1}(\frac{1}{2} r^2) e^{-\frac{1}{4} r^2}$.

Proof: Since the operator $f \to f \times \mu_r$ is easily seen to be bounded on $L^2(\mathbb{C}^n)$ and

$$f = (2\pi)^{-n} \sum_{k=0}^{\infty} f \times \varphi_k(z),$$

it is enough to show that

$$f \times \varphi_k \times \mu_r(z) = \frac{k!(n-1)!}{(k+n-1)!} \varphi_k(r) f \times \varphi_k(z).$$

As we already know, the functions

$$\psi_k(r) = \left(\frac{2^{1-n}k!}{(k+n-1)!} \right)^{\frac{1}{2}} \varphi_k(r)$$

form an orthonormal basis for $L^2(R_+, r^{2n-1}dr)$ and so $\varphi_k \times \mu_r(z)$, considered as a function of r, can be expanded in terms of $\psi_k(r)$. A simple calculation shows that

$$\int_0^\infty \varphi_k \times \mu_r(z)\psi_j(r)r^{2n-1}\, dr =$$

$$(2\pi)^{-n}2^{n-1}(n-1)! \left(\frac{2^{1-n}k!}{(k+n-1)!} \right)^{\frac{1}{2}} \varphi_k \times \varphi_j(z).$$

In view of the orthogonality relations

$$\varphi_k \times \varphi_j(z) = (2\pi)^n \delta_{kj} \varphi_k(z)$$

it follows that

$$\varphi_k \times \mu_r(z) = \frac{k!(n-1)!}{(k+n-1)!}\varphi_k(r)\varphi_k(z),$$

and this proves the theorem. ∎

Finally, we are in a position to prove our main result. As we have already remarked, it is enough to prove the following theorem for functions on $\mathbb{C}$. We let $f_m(z) = z^m f(z)$ or $\bar{z}^m f(z)$ according to whether $m \geq 0$ or $m \leq 0$ and

$$\psi_m(k, z) = f_m \times \varphi_k(z)$$

be the special Hermite projections of f_m.

Theorem 2.4.5 *Let f be a continuous function on $\mathbb{C}$. Then f is compactly supported in $|z| \leq B$ if and only if*

$$\|\psi_m\|_{2,j} \leq C_m 2^{-j}(B + |z|)^{2j}$$

for all $j \geq 0$.

Proof: Consider the function $f \times \mu_r(z)$. If f is supported in $|z| \leq B$ then it follows that $f \times \mu_r(z) = 0$ for $r > (B + |z|)$. In other words, the radial function

$$u_z(\zeta) = f \times \mu_\zeta(z)$$

is supported in $|\zeta| \leq (B + |z|)$ for each z. Note that

$$u_z(\zeta) = (2\pi)^{-1} \sum_{k=0}^{\infty} f \times \varphi_k(z) \varphi_k(\zeta)$$

and applying Theorem 2.4.3 to this radial function, we get

$$\|\psi_0\|_{2,j} \leq C 2^{-j} (B + |z|)^{2j}$$

for all j. A similar argument applies to all f_m and this proves the direct part of the theorem.

   To prove the converse, first assume that f is radial. If $\psi_0(z) = (f \times \varphi_k(z))$ satisfies the conditions of the theorem then $u_z(\zeta)$ is supported in $|\zeta| \leq (B + |z|)$. As f is radial

$$f \times \varphi_k(z) = \left(\int_C f(z) \varphi_k(z) \, dz \right) \varphi_k(z)$$

and therefore, $u_z(\zeta) = u_\zeta(z)$. This means that $f \times \mu_r(z)$ vanishes for $|z| > (B + r)$. Letting r tend to zero, we conclude that f vanishes for $|z| > B$.

   If f is not radial consider its radialisation

$$F(z) = \int_0^{2\pi} f(e^{i\theta} z) \, d\theta.$$

A calculation shows that

$$F \times \varphi_k(z) = \int_0^{2\pi} f \times \varphi_k(e^{i\theta} z) \, d\theta$$

and therefore, if we let $\Psi(k, z) = F \times \varphi_k(z)$ then

$$\Delta^j \Psi(k, z) = \int_0^{2\pi} \Delta^j \psi_0(k, e^{i\theta} z) \, d\theta.$$

This shows that

$$\|\Delta^j \Psi(k, z)\|_{2,j} \leq C 2^{-j} (B + |z|)^{2j}$$

and as F is radial, we conclude that F is supported in $|z| \leq B$.

The same argument applied to the radialisation F_m of f_m shows that F_m is supported in $|z| \leq B$. But when $m \geq 0$, we have

$$F_m(z) = \int_0^{2\pi} (e^{i\theta} z)^m f(e^{i\theta} z) \, d\theta$$

and this shows that

$$F_m(r) = r^m \int_0^{2\pi} e^{mi\theta} f(e^{i\theta} r) \, d\theta$$

is nothing but the $(-m)$th Fourier coefficient of the function $\theta \to f(e^{i\theta} z)$. By considering $m \leq 0$ as well, we conclude that all the Fourier coefficients of the function $f(e^{i\theta} z)$ vanish when $|z| > B$, which means that f is supported in $|z| \leq B$. This completes the proof of the theorem. ∎

2.5   Bochner-Riesz means for the sublaplacian

For L^2 functions the inversion formula for the Euclidean Fourier transform on $\mathbb{R}^n$ takes the form

$$f(x) = \lim_{R \to \infty} (2\pi)^{-\frac{n}{2}} \int_{|\xi| \leq R} e^{ix \cdot \xi} \hat{f}(\xi) \, d\xi$$

where the limit exists in the L^2 norm. The right hand side defines the partial sum operator $S_R f$ and the convergence of $S_R f$ to f in the L^2 norm is equivalent to the uniform boundedness of $S_R f$ on $L^2(\mathbb{R}^n)$. As shown by Fefferman [22], the partial sum operator is not uniformly bounded on any other L^p spaces. Therefore, if we want an inversion formula for the Fourier transform on L^p spaces, we have to introduce some summability factors. One such summability method is provided by the Bochner-Riesz means, defined by

$$S_R^\alpha f(x) = (2\pi)^{-\frac{n}{2}} \int_{|\xi| \leq R} e^{ix \cdot \xi} \left(1 - \frac{|\xi|^2}{R^2}\right)^\alpha \hat{f}(\xi) \, d\xi.$$

There is an extensive literature concerning the boundedness properties of these operators; see Stein [65], Sogge [62] and Davis-Chang [15]. The Bochner-Riesz means are convolution operators and when $\alpha > \frac{(n-1)}{2}$ they are uniformly bounded on L^p for all $1 \leq p \leq \infty$. When $0 < \alpha \leq \frac{(n-1)}{2}$, the well known Bochner-Riesz conjecture says that S_R^α are uniformly bounded on L^p for all p satisfying $\frac{2n}{n+1+2\alpha} < p < \frac{2n}{n-1-2\alpha}$. The conjecture is completely settled only when $n = 2$ (see Carleson-Sjölin [11]) but in higher dimensions only partial results are known.

In this section we define and study the Bochner-Riesz means $S_R^\alpha f$ associated to the sublaplacian on $L^p(H^n)$. The sublaplacian $\mathcal{L}$ is a formally nonnegative hypoelliptic differential operator which has a unique self adjoint extension to $L^2(H^n)$. Let $E(\lambda)$ be the spectral resolution of this extension so that we have

$$\mathcal{L}f = \int_0^\infty \lambda \, dE(\lambda)f.$$

The Bochner-Riesz means $S_R^\alpha f$ associated to the sublaplacian is then defined by

$$S_R^\alpha f = \int_0^R \left(1 - \frac{\lambda}{R}\right)^\alpha dE(\lambda)f.$$

If we let

$$\varphi_R^\alpha(\lambda) = \left(1 - \frac{\lambda}{R}\right)_+^\alpha,$$

where $\left(1 - \frac{\lambda}{R}\right)_+$ is the positive part of $\left(1 - \frac{\lambda}{R}\right)$, then

$$S_R^\alpha f = \varphi_R^\alpha(\mathcal{L})f.$$

In terms of the spectral projections $\mathcal{P}_\lambda f$ studied in the previous sections we have

$$dE(\lambda)f = \mathcal{P}_\lambda f d\mu(\lambda)$$

so that

$$S_R^\alpha f = \int_0^R \left(1 - \frac{\lambda}{R}\right)^\alpha \mathcal{P}_\lambda f d\mu(\lambda).$$

The boundedness of S_R^α on L^p spaces has been investigated by several authors; see the references given in Section 2.7.

In this section we establish the following theorem. Let

$$\alpha(p) = (2n + 1)\left|\frac{1}{p} - \frac{1}{2}\right|, \quad 1 \le p \le \infty.$$

Theorem 2.5.1 *Assume that $1 \le p \le \infty$ and $\alpha > \alpha(p)$. Then the uniform estimates*

$$\|S_R^\alpha f\|_p \le C\|f\|_p$$

hold for all $f \in L^p(H^n)$. If $p < \infty$, then $S_R^\alpha f$ converges to f in the L^p norm as R tends to infinity.

In proving this theorem we follow the well known method of using the restriction theorem for the projections $E_{a,b}$ studied in Section 2.3. This method has been applied in various setups to study the Bochner-Riesz means; see for example Sogge [62] and [84]. In order to carry out this method of proof we need good pointwise estimates for the kernel of the Riesz means S_R^α when α is large. In order to get this we use the Hausdorff-Young inequality for the inverse Fourier transform on H^n applied to radial functions.

As the projections $\mathcal{P}_\lambda f$ are convolution operators with kernels $G_\lambda(z,t)$ which we know explicitly, the Bochner-Riesz means take the form

$$S_R^\alpha f(z,t) = \sum_{k=0}^\infty \int_{-\infty}^\infty \left(1 - \frac{(2k+n)|\lambda|}{R}\right)_+^\alpha f * e_k^\lambda(z,t)\, d\mu(\lambda).$$

Thus $S_R^\alpha f = f * S_R^\alpha$ where the kernel $S_R^\alpha(z,t)$ is given by

$$S_R^\alpha(z,t) = \sum_{k=0}^\infty \int_{-\infty}^\infty e^{-i\lambda t} \left(1 - \frac{(2k+n)|\lambda|}{R}\right)_+^\alpha \varphi_k^\lambda(z)\, d\mu(\lambda).$$

From the above expression it is clear that

$$S_R^\alpha(z,t) = R^{n+1} S_1^\alpha(\sqrt{R}z, Rt).$$

Therefore, in order to prove the theorem it is enough to show that $S_1^\alpha f$ is bounded on $L^p(H^n)$ whenever $\alpha > \alpha(p)$. From the explicit form of $S_1^\alpha(z,t)$ it is clear that it is radial.

When F is radial we know that its Fourier transform is given by

$$\hat{F}(\lambda) = \sum_{k=0}^\infty R_k(\lambda, F) P_k(\lambda)$$

where $P_k(\lambda)$ are the projections associated to

$$H(\lambda) = -\Delta + \lambda^2 |x|^2$$

and $R_k(\lambda, F)$ are the Laguerre coefficients of F^λ given by

$$R_k(\lambda, F) = \frac{2^{(1-n)}k!}{(k+n-1)!} \int_0^\infty F^\lambda(r) \varphi_k^\lambda(r)\, r^{2n-1}\, dr.$$

The inversion formula for the Fourier transform then reads

$$F(z,t) = \sum_{k=0}^\infty \int_{-\infty}^\infty e^{-i\lambda t} R_k(\lambda, F) \varphi_k^\lambda(z)\, d\mu(\lambda)$$

and the Plancherel theorem takes the form

$$\|f\|_2^2 = c_n \int_{-\infty}^{\infty} \left(\sum_{k=0}^{\infty} |R_k(\lambda, F)|^2 \frac{(k+n-1)!}{k!} \right) d\mu(\lambda).$$

If we start with a sequence of measurable functions $R_k(\lambda)$ such that

$$\int_{-\infty}^{\infty} \left(\sum_{k=0}^{\infty} |R_k(\lambda)| \frac{(k+n-1)!}{k!} \right) d\mu(\lambda)$$

is finite, then the function F defined by

$$F(z,t) = \sum_{k=0}^{\infty} \int_{-\infty}^{\infty} e^{-i\lambda t} R_k(\lambda) \varphi_k^\lambda(z) \, d\mu(\lambda) \qquad (2.5.7)$$

is in L^∞. This follows from the fact that

$$\|\varphi_k\|_\infty = c_n \frac{(k+n-1)!}{k!}.$$

Interpolating with the Plancherel theorem, we obtain the following version of the Hausdorff-Young inequality for the inverse Fourier transform.

Proposition 2.5.2 *Let $1 \le p \le 2$ and let $\{R_k(\lambda)\}$ be a sequence of measurable functions such that*

$$\|R_k(\lambda)\|_p = \int_{-\infty}^{\infty} \left(\sum_{k=0}^{\infty} |R_k(\lambda)|^p \frac{(k+n-1)!}{k!} \right) d\mu(\lambda)$$

is finite. Then the function F defined in (2.5.7) belongs to $L^{p'}$ and we have the inequality

$$\|F\|_{p'} \le C\|R_k(\lambda)\|_p.$$

Moreover, $R_k(\lambda, F) = R_k(\lambda)$.

Using the proposition with $p = 1$, we now establish the following estimate for the kernel of the Riesz means. For $w = (z,t) \in H^n$ define $|w|^4 = |z|^4 + t^2$. This defines a norm on H^n which is homogeneous of degree 1 with respect to the nonisotropic dilations.

Theorem 2.5.3 *Assume that $\alpha > 2m$ where m is a positive integer. Then*

$$|S_R^\alpha(w)| \le CR^{(n+1)} \left(1 + R^{\frac{1}{2}}|w| \right)^{-2m}$$

for all $w = (z,t) \in H^n$.

Proof: As we have remarked earlier, it is enough to get the estimate when $R = 1$. We let $F(w) = S_1^\alpha(w)$ and calculate

$$R_k(\lambda, (it - \frac{1}{4}|z|^2)^m F)$$

in terms of $R_k(\lambda, F)$. If we could show that

$$\int_{-\infty}^{\infty} \sum_{k=0}^{\infty} |R_k(\lambda, (it - \frac{1}{4}|z|^2)^m F)| \frac{(k + n - 1)!}{k!} d\mu(\lambda)$$

is finite, then from the proposition it follows that

$$(it - \frac{1}{4}|z|^2)^m F(z, t)$$

is bounded and the theorem is proved. To calculate

$$R_k(\lambda, (it - \frac{1}{4}|z|^2)^m F)$$

we write $r = |z|$ so that

$$F(r, t) = \sum_{k=0}^{\infty} \int_{-\infty}^{\infty} e^{-i\lambda t} (1 - (2k + n)|\lambda|)_+^\alpha \varphi_k^\lambda(r) \, d\mu(\lambda).$$

This means that

$$R_k(\lambda, F) = (1 - (2k + n)|\lambda|)_+^\alpha .$$

We first calculate $R_k(\lambda, (it - \frac{1}{4}|z|^2)F)$. Recalling the definition of $R_k(\lambda, F)$ we have

$$R_k(\lambda, itF) = \frac{2^{(1-n)}k!}{(k + n - 1)!} \int_0^{\infty} (itF)^\lambda(r)\varphi_k^\lambda(r)r^{2n-1} \, dr.$$

Since $(itF)^\lambda(r) = \frac{d}{d\lambda} F^\lambda(r)$ we have the equation

$$R_k(\lambda, itF) =$$

$$\frac{d}{d\lambda} R_k(\lambda, F) - \frac{2^{(1-n)}k!}{(k + n - 1)!} \int_0^{\infty} F^\lambda(r) \frac{d}{d\lambda} \varphi_k^\lambda(r)r^{2n-1} \, dr.$$

Now assuming $\lambda > 0$ for a moment

$$\frac{d}{d\lambda} \varphi_k^\lambda(r)$$

$$= \frac{1}{2}r^2 \frac{d}{dr}L_k^{n-1}(\frac{1}{2}\lambda r^2)e^{-\frac{\lambda}{4}r^2} - \frac{1}{4}r^2 L_k^{n-1}(\frac{1}{2}\lambda r^2)e^{-\frac{\lambda}{4}r^2}.$$

Using the recursion formula (see 5.1.14 of Szego [72])

$$r\frac{d}{dr}L_k^{n-1}(r) = kL_k^{n-1}(r) - (k+n-1)L_{k-1}^{n-1}(r)$$

we can write the above as

$$\frac{d}{d\lambda}\varphi_k^\lambda(r)$$

$$= \lambda^{-1}\left(k\varphi_k^\lambda(r) - (k+n-1)\varphi_{k-1}^\lambda(r)\right) - \frac{1}{4}r^2\varphi_k^\lambda(r)$$

and consequently we have proved the formula

$$R_k(\lambda, (it - \frac{1}{4}r^2)F) =$$

$$\frac{d}{d\lambda}R_k(\lambda, F) - k\lambda^{-1}\left(R_k(\lambda, F) - R_{k-1}(\lambda, F)\right).$$

When $\lambda < 0$ we get a similar formula where $k\lambda^{-1}$ on the right hand side of the above equation is replaced by $k|\lambda|^{-1}$.

Now we let $\sigma = (2k+n)|\lambda|$ so that $R_k(\lambda, F) = \psi(\sigma)$ where

$$\psi(\sigma) = (1-\sigma)_+^\alpha .$$

Defining

$$\psi_1(\sigma) = R_k(\lambda, (it - \frac{1}{4}r^2)F),$$

the formula can be written in the form

$$\psi_1(\sigma) = |\lambda|^{-1}\{\sigma\psi'(\sigma) - k\psi(\sigma) + k\psi(\sigma - 2|\lambda|)\}.$$

The function $\psi(\sigma)$ has the following two properties: (i) it is supported in a set of the form $(2k+n)|\lambda| \leq c$, for large k and (ii)

$$\int |\psi((2k+n)|\lambda|)| \, d\mu(\lambda) \leq C(2k+n)^{-n-1}.$$

We claim that the function $\psi_1(\sigma)$ also satisfies the same conditions. From the above expression for $\psi_1(\sigma)$ in terms of $\psi(\sigma)$ it is clear that it has the property (i).

To see that $\psi_1(\sigma)$ also has the property (ii), we rewrite $\psi_1(\sigma)$ as

$$\psi_1(\sigma) = \frac{n}{2}|\lambda|^{-1}\frac{\partial}{\partial k}\psi((2k+n)|\lambda|)$$

$$+2k|\lambda|^{-1}\{\frac{\partial}{\partial k}\psi((2k+n)|\lambda|) - \psi((2k+n)|\lambda|) + \psi((2k-2+n)|\lambda|)\}.$$

Using Taylor expansion we can write

$$\frac{\partial}{\partial k}\psi((2k+n)|\lambda|) - \psi((2k+n)|\lambda|) + \psi((2k-2+n)|\lambda|)$$

$$= 4|\lambda|^2 \int_{k-1}^{k} (t+1-k)\psi''((2t+n)|\lambda|)\, dt.$$

Therefore, the integral of

$$2k|\lambda|^{-1}\{\frac{\partial}{\partial k}\psi((2k+n)|\lambda|) - \psi((2k+n)|\lambda|) + \psi((2k-2+n)|\lambda|)\}$$

with respect to $d\mu(\lambda)$ is bounded by

$$C\int_{k-1}^{k} dt \int (1-(2t+n)|\lambda|)_+^{\alpha-2}\, d\mu(\lambda)$$

which in turn is bounded by $(2k+n)^{-n-1}$. We can also show that the integral of

$$\frac{n}{2}|\lambda|^{-1}\frac{\partial}{\partial k}\psi((2k+n)|\lambda|)$$

has the same estimate. This proves that $\psi_1(\sigma)$ also has the property (ii).

Now an iteration of the process shows that

$$R_k\left(\lambda, (it-\frac{1}{4}r^2)^j F\right) = \psi_j(\sigma)$$

satisfies the condition

$$\int |\psi_j((2k+n)|\lambda|)|\, d\mu(\lambda) \leq C_j(2k+n)^{-n-1}$$

provided $(1-\lambda)_+^{\alpha-2}$ is integrable. Hence when $j = m$ and $\alpha > 2m$ we get

$$\int |R_k\left(\lambda, (it-\frac{1}{4}r^2)^m F\right)|\, d\mu(\lambda) \leq C_m(2k+n)^{-n-1}$$

and consequently

$$\sum_{k=0}^{\infty} \int |R_k\left(\lambda, (it-\frac{1}{4}r^2)^m F\right)|\, d\mu(\lambda)\frac{(k+n-1)!}{k!} \leq C_m.$$

This completes the proof of the theorem.   ■

    To prove the main result of this section we take a partition of unity

$$\sum_{-\infty}^{\infty} \varphi(2^j s) = 1$$

where $\varphi \in C_0^{\infty}(\frac{1}{2}, 2)$. Let us write

$$\varphi_j^{\alpha}(s) = (1 - s)^{\alpha} \varphi(2^j (1 - s))$$

and define $T_j = \varphi_j^{\alpha}(\mathcal{L})$. Then $S_1^{\alpha} f = \sum_{j=0}^{\infty} T_j$ and our result on the Bochner-Riesz means follows once we prove the following result.

Proposition 2.5.4 *Assume that* $\alpha > \alpha(p)$. *Then there is an* $\epsilon > 0$ *such that*

$$\|T_j f\|_p \le C 2^{-\epsilon j} \|f\|_p$$

for all $f \in L^p(H^n)$.

In order to prove the proposition we split the kernel $s_j^{\alpha}(w)$ of T_j into two parts:

$$s_j^{\alpha}(w) = K_j^1(w) + K_j^2(w)$$

where

$$K_j^1(w) = s_j^{\alpha}(w) \chi(|w| \le 2^{j(1+\gamma)})$$

and $K_j^2(w) = s_j^{\alpha}(w) - K_j^1(w)$. Here $|w|$ is the homogeneous norm and $\chi(A)$ stands for the characteristic function of A and $\gamma > 0$ is to be fixed later. First we treat the convolution with $K_j^2(w)$.

Proposition 2.5.5 *Given any* $\gamma > 0$ *there is an* $\epsilon > 0$ *such that*

$$\|f * K_j^2\|_p \le C 2^{-\epsilon j} \|f\|_p.$$

Proof: It is here in the proof of this proposition we need the kernel estimates proved in Theorem 2.5.3. We will show that there is an $\epsilon > 0$ such that

$$\int_{H^n} |K_j^2(w)| \, dw \le C 2^{-\epsilon j}$$

which will immediately prove the proposition. In order to prove the above L^1 estimate for the kernel K_j^2, we need to recall a few elementary facts concerning Riesz means.

If we set $S(t,w) = S_t^0(w)$ to be the kernel of the partial sum operator, then

$$t \to S(t,w)$$

is a function of bounded variation and

$$s_j^\alpha(w) = -\int \varphi_j^\alpha(t) \, dS(t,w).$$

Integrating by parts and making use of the identity

$$\frac{\partial}{\partial t}(t^m S_t^m(w)) = mt^{m-1}S_t^{m-1}(w)$$

we get the relation

$$s_j^\alpha(w) = c_m \int t^{2m+1} S_t^{2m+1}(w) \left(\frac{\partial}{\partial t}\right)^{2m+2}(\varphi_j^\alpha(t)) \, dt.$$

Since φ_j^α is supported in $2^{-j-1} \le (1-t) \le 2^{-j+1}$ we have the bound

$$\left|\left(\frac{\partial}{\partial t}\right)^{2m+2}(\varphi_j^\alpha(t))\right| \le 2^{(2m+2)j}.$$

The estimates of Theorem 2.5.3, together with the above estimate, gives us

$$|s_j^\alpha(w)| \le C2^{(2m+2)j}|w|^{-2m}.$$

From this estimate it follows that

$$\int_{|w| \ge 2^{j(1+\gamma)}} |s_j^\alpha(w)| \, dw$$

$$\le C2^{(2m+2)j} \int_{2^{j(1+\gamma)}}^\infty t^{-2m+Q-1} \, dt$$

which is bounded by $C2^{(2m+2)j}2^{j(1+\gamma)(-2m+Q)}$. Choosing m so large that

$$(2m+2)\gamma > (Q+2)(\gamma+1)$$

we can obtain the estimate

$$\int_{H^n} |K_j^2(w)| \, dw \le C2^{-\epsilon j}$$

for some ϵ and the proposition follows. ∎

Since we are assuming $\alpha > (Q-1)|\frac{1}{p} - \frac{1}{2}|$ we can choose $\gamma > 0$ so that

$$\alpha > (Q(1+\gamma) - 1)|\frac{1}{p} - \frac{1}{2}|.$$

Using this γ we split the function f into three parts: $f = f_1 + f_2 + f_3$. For $\zeta \in H^n$ let $f_1(w) = f(w)$ when $|w - \zeta| \leq \frac{3}{4}2^{j(1+\gamma)}$ and $f_1(w) = 0$ otherwise; $f_2(w) = f(w)$ when $\frac{3}{4} 2^{j(1+\gamma)} \leq |w - \zeta| \leq \frac{5}{4} 2^{j(1+\gamma)}$ and $f_2(w) = 0$ otherwise; and $f_3 = f - f_1 - f_2$. In order to prove Proposition 2.5.4 it remains to consider convolution with the kernel $K_j^1(w)$. Let $B(\zeta, r)$ stand for the ball $|w - \zeta| \leq r2^{j(1+\gamma)}$. We will show that

$$\int_{B(\zeta,\frac{1}{4})} |K_j^1 f(w)|^p \, dw \leq C2^{-\epsilon jp} \int_{B(\zeta,\frac{5}{4})} |f(w)|^p \, dw.$$

Integration with respect to ζ will then show that

$$\int_{H^n} |K_j^1 f(w)|^p \, dw \leq C2^{-\epsilon jp} \int_{H^n} |f(w)|^p \, dw.$$

This together with Proposition 2.5.5 will then prove Proposition 2.5.4. When

$$|w - \zeta| \leq \frac{1}{4}2^{j(1+\gamma)}$$

and w' belongs to the support of f_3 it follows that

$$|w - w'| > 2^{j(1+\gamma)}$$

and consequently $K_j^1 f_3 = 0$. When

$$|w - \zeta| \leq \frac{1}{4}2^{j(1+\gamma)}$$

and w' belongs to the support of f_2 we have

$$|w - \zeta| > \frac{1}{2}2^{j(1+\gamma)}$$

and so we can repeat the proof of Proposition 2.5.5 to conclude that

$$\int_{B(\zeta,\frac{1}{4})} |K_j^1 f_2(w)|^p \, dw \leq C2^{-\epsilon jp} \int_{B(\zeta,\frac{5}{4})} |f(w)|^p \, dw.$$

Finally, $K_j^1 f_1$ will be taken care of once we prove the following proposition.

Proposition 2.5.6 *For any ball B of radius $2^{j(1+\gamma)}$ we have*

$$\|f * s_j^\alpha\|_{L^p(B)} \le C2^{-\epsilon j}\|f\|_p.$$

Proof: Applying Hölder's inequality

$$\|f * s_j^\alpha\|_{L^p(B)} \le |B|^{(\frac{1}{p}-\frac{1}{2})}\|f * s_j^\alpha\|_2$$

where $|B|$ stands for the volume of B. Since

$$f * s_j^\alpha = \varphi_j^\alpha(\mathcal{L})f$$

we have

$$f * s_j^\alpha = \int_{1-2^{-j+1}}^{1-2^{-j-1}} (1-\lambda)^\alpha \mathcal{P}_\lambda f \, d\mu(\lambda).$$

We can adapt the proof of the restriction theorem for $E_{a,b}$ to conclude that

$$\|f * s_j^\alpha\|_2 \le C2^{-\alpha j}2^{-j(\frac{1}{p}-\frac{1}{2})}\|f\|_p.$$

Since $|B| = c_n 2^{j(1+\gamma)Q}$ we finally get

$$\|f * s_j^\alpha\|_{L^p(B)} \le C2^{-j\alpha}2^{(Q(1+\gamma)-1)(\frac{1}{p}-\frac{1}{2})}\|f\|_p.$$

By the choice of γ it follows that

$$\|f * s_j^\alpha\|_{L^p(B)} \le C2^{-j\epsilon}\|f\|_p$$

for some $\epsilon > 0$ and this proves the proposition. ∎

Thus we have shown that the Bochner-Riesz means are uniformly bounded. For Schwartz class functions we can show that they actually converge to the function in the norm. By the usual density argument we can establish the convergence of the Bochner-Riesz means to the function in the norm. This completes the proof of Theorem 2.5.1. Using the mixed norm estimates for the projections $E_{a,b}$, we can get mixed norm estimates for the Bochner-Riesz means, see Müller [45]. We have shown that when $\alpha > \frac{(Q-1)}{2}$, the operator S_1^α is bounded on $L^1(H^n)$. It is not known whether this condition is necessary. Also when $\alpha \le \frac{(Q-1)}{2}$ nothing is known about the boundedness of S_1^α. Probably a condition like $\alpha > Q(\frac{1}{p} - \frac{1}{2}) - \frac{1}{2}$ is necessary and sufficient for the boundedness of S_1^α on $L^p(H^n)$, but this is still an open problem. We only know that $\alpha > 2n(\frac{1}{p} - \frac{1}{2}) - \frac{1}{2}$ is necessary for uniform boundedness; see [45].

2.6   A multiplier theorem for the Fourier transform

For the Euclidean Fourier transform on $\mathbb{R}^n$, we have the multiplier transform T_m defined by

$$(T_m f\hat{)}(\xi) = m(\xi)\hat{f}(\xi).$$

Here $m(\xi)$ is a bounded function called the multiplier. Some basic examples of multiplier transforms are provided by the Riesz transforms and the Bochner-Riesz means. The multiplier transform T_m is clearly bounded on $L^2(\mathbb{R}^n)$, but without further assumptions on m, it does not have to define a bounded operator on $L^p(\mathbb{R}^n)$ for $p \neq 2$. A sufficient condition on $m(\xi)$ is given by Hörmander [33] and Mihlin [43] in terms of its derivatives.

In this section we study such multiplier transforms for the group Fourier transform on the Heisenberg group. As the Fourier transform on H^n is operator valued, the analogue of the multiplier transform in the setting of the Heisenberg group is the following: if $M(\lambda)$ is an operator valued function, then we can define

$$(T_M f\hat{)}(\lambda) = \hat{f}(\lambda)M(\lambda)$$

and, as in the case of $\mathbb{R}^n$, we can ask under what conditions on $M(\lambda)$ the above operator T_M initially defined on the space of Schwartz class functions extends to $L^p(H^n)$ as a bounded operator.

Such multiplier transformations have been studied by several authors in the literature. The Bochner-Riesz means studied in the previous section is an example of a multiplier transform. In this section we consider multipliers which are of some particular form. Recall that the scaled Hermite operator $H(\lambda)$ and the sublaplacian $\mathcal{L}$ are related by

$$(\mathcal{L}f\hat{)}(\lambda) = \hat{f}(\lambda)H(\lambda).$$

Given a bounded function m defined on $(0, \infty)$, we consider multipliers of the form $M(\lambda) = m(H(\lambda))$. In terms of the sublaplacian this means that $T_M f = m(\mathcal{L})f$. This corresponds to the radial multipliers $m(|\xi|^2)$ in the case of $\mathbb{R}^n$ because then $T_m f = m(-\Delta)f$. In this section we find sufficient conditions on the function m so that $m(\mathcal{L})$ extends to a bounded operator on $L^p(H^n)$. The following is the analogue of the radial version of the Hörmander-Mihlin multiplier theorem.

Theorem 2.6.1 *Let m be an $(n + 1)$ times differentiable function and*

$$|m^{(j)}(t)| \leq C_j t^{-j}$$

for $j = 0, 1, \ldots, (n+1)$. *Then* $m(\mathcal{L})$ *extends to a bounded operator on* $L^p(H^n)$ *for* $1 < p < \infty$.

This result is due to Müller and Stein [46]. A slightly weaker version of the above theorem which involves the same conditions on the derivatives of m up to order $n+2$ was known for some time. As $n+2$ is the smallest integer greater than $\frac{Q}{2}$ where Q is the homogeneous dimension of H^n, it was expected that the condition involving $n+2$ derivatives was the optimal one. So, the above theorem came as a surprise, at least to the author. The proof of the above theorem is a bit long and involved; to keep this exposition simple we only prove the previously known weaker version. To prove the multiplier theorem we use the Littlewood-Paley-Stein theory of $g-$functions for the semigroup generated by the sublaplacian.

In [63], Stein developed the Littlewood-Paley theory for general contraction semigroups satisfying certain conditions and applied the $g-$function estimates to prove a universal multiplier theorem. Again in [63], Stein used the same method to give a different proof of the Hörmander-Mihlin multiplier theorem. This method turned out to be very useful in other settings as well; see Strichartz [68] for the case of spherical harmonics, [76] for the Hermite expansions and [77] for the Weyl transform. In this section we develop the Littlewood-Paley-Stein theory for the semigroup generated by the sublaplacian and use it to prove a weaker version of Theorem 2.6.1. Let T^s stand for the semigroup generated by $\mathcal{L}$. The following result of Folland [25] summarises the properties of this semigroup.

Theorem 2.6.2 *The sublaplacian generates a unique semigroup* T^s, $s > 0$ *of linear operators on* $L^1 + L^\infty$ *satisfying the following conditions: (i)* $T^s f = f * h_s$ *where* $h_s(w) = h(s, w)$ *is a* C^∞ *function on* $H^n \times (0, \infty)$, $\int h_s(w) \, dw = 1$ *for all* s *and* $h_s(w) \geq 0$ *for all* w *and* s. *(ii)* T^s *is a contraction semigroup on* $L^p(H^n)$, $1 \leq p \leq \infty$ *which is strongly continuous for* $p < \infty$. *(iii)* T^s *is self adjoint,* $f \geq 0$ *implies* $T^s f \geq 0$ *and* $T^s 1 = 1$.

Proof: Let $\tilde{D}$ be the space of C^∞ functions which are constant outside a compact set. Let D be the completion of $\tilde{D}$ with respect to the uniform norm and let D^2 be the completion of D with respect to the norm

$$\|f\|_\infty + \sum_{j=1}^{2n+1} \|Y_j f\|_\infty + \sum_{j=1}^{2n+1} \sum_{k=1}^{2n+1} \|Y_j Y_k f\|_\infty$$

where Y_j forms a basis for the Heisenberg Lie algebra. Then according to a theorem of G. Hunt there is a unique strongly continuous semigroup T^s on $\tilde{D}$ such that (a) for each $s > 0$ there is a probability measure μ_s on H^n such that

$$T^s f(w) = \int_{H^n} f(wv^{-1}) \, d\mu_s(v)$$

(b) the infinitesimal generator of T^s is defined on D^2 and there it coincides with $\mathcal{L}$. Moreover,

$$\lim_{s \to 0} \mu_s(E) = 1$$

whenever E contains 0 and as $\mathcal{L}$ is symmetric $d\mu_s(v) = d\mu_s(v^{-1})$. We also note that since $\mathcal{L}$ annihilates constants and $\tilde{D}$ is dense in D^2, the action of T^s on $\tilde{D}$ determines T^s.

Let h be the distribution on $H^n \times (0, \infty)$ defined by

$$(h, u.v) = \int_0^\infty \int_{H^n} u(w)v(s) \, d\mu_s(w)$$

where $u \in C_0^\infty(H^n)$ and $v \in C_0^\infty(0, \infty)$. Then because of (b) it is clear that

$$(h, \mathcal{L}u.v) = (h, u.\frac{dv}{ds})$$

so that h is a distribution solution of

$$(\mathcal{L} + \frac{\partial}{\partial s})h = 0.$$

But by a theorem of Hörmander, $\mathcal{L} + \frac{\partial}{\partial s}$ is hypoelliptic and so $h \in C^\infty(H^n \times (0, \infty))$ and we have $d\mu_s(w) = h(w,s)dw$. Thus $h_s(w) \geq 0, \int h_s(w) \, dw = 1$ and T^s is self adjoint since $h_s(w) = h_s(w^{-1})$. By (i) and Young's inequality T^s is a contraction semigroup on $L^p(H^n)$, $1 \leq p \leq \infty$; it is strongly continous for $p < \infty$ as $h_s \to \delta$ as s tends to zero. ∎

Associated to the semigroup T^s are the following g and g^* functions. For each positive integer k we define

$$g_k(f, w)^2 = \int_0^\infty |\partial_s^k T^s f(w)|^2 s^{2k-1} \, ds,$$

$$g_k^*(f, w)^2 = \int_0^\infty \int_{H^n} s^{-n} \left(1 + s^{-2}|v|^4\right)^{-k} |\partial_s^k T^s f(v^{-1}w)|^2 \, ds dv.$$

For these operators we have the following result.

Theorem 2.6.3 *(i) For $k \geq 1$,*

$$\|g_k(f)\|_2 = 2^{-k}\|f\|_2$$

(ii) For $1 < p < \infty$, there are constants C_1 and C_2 such that

$$C_1\|f\|_p \leq \|g_k(f)\|_p \leq C_2\|f\|_p$$

(iii) If $k > \frac{n+1}{2}$ and $p > 2$ then

$$\|g_k^*(f)\|_p \leq C\|f\|_p.$$

Proof: The inequality

$$\|g_k(f)\|_p \leq C_2\|f\|_p$$

follows from the general theory of g functions for contraction semigroups; see Stein [63]. The reverse inequality can be easily deduced once we have (i). When $k > \frac{n+1}{2}$ the function $(1 + s^{-2}|v|^4)^{-k}$ is integrable and hence we can prove (iii) using (i). These proofs are standard and we omit the details. However, we give a proof of (i) using the Plancherel theorem for the Fourier transform on H^n.

We assume $k = 1$, the general case being similar. From the definition it follows that

$$\|g_1(f)\|_2^2 = \int_0^\infty \int_{H^n} |\partial_s T^s f(w)|^2 s \, dw \, ds.$$

Applying the Plancherel theorem to the inner integral, we get

$$\|g_1(f)\|_2^2 = \int_0^\infty \int_{-\infty}^\infty \|(\partial_s T^s f\hat{)}(\lambda)\|_{HS}^2 \, d\mu(\lambda) \, ds.$$

But we have

$$(\partial_s T^s f\hat{)}(\lambda) = -\hat{f}(\lambda)H(\lambda)e^{-sH(\lambda)}$$

and therefore, its squared Hilbert-Schmidt norm is given by the sum

$$\sum \left((2|\alpha| + n)|\lambda|\right)^2 e^{-2s(2|\alpha|+n)|\lambda|}\|\hat{f}(\lambda)\Phi_\alpha^\lambda\|_2^2.$$

Integrating this with respect to sds, we get

$$\|g_1(f)\|_2^2 = \frac{1}{4}\int_{-\infty}^\infty \|\hat{f}(\lambda)\|_{HS}^2 \, d\mu(\lambda) = \frac{1}{4}\|f\|_2^2$$

which proves the asserted estimate on $g_1(f)$. ∎

We now start the proof of (a weaker version of) the multiplier theorem. Slightly abusing the notation, let us write $Mf = m(\mathcal{L})f$. In view of the previous theorem, it is enough to show that the pointwise estimate

$$g_{k+1}(Mf, w) \leq Cg_k^*(f, w) \qquad (2.6.8)$$

holds for some integer $k > \frac{n+1}{2}$. Then the multiplier theorem for $p > 2$ will follow immediately from the previous theorem. For $1 < p < 2$, we can use a duality argument.

So we proceed to prove the estimate (2.6.8). Let us write

$$u_s = T^s f, \quad U_s = T^s(Mf).$$

Then it is easily verified that

$$U_{t+s}(w) = G_t * u_s(w) \qquad (2.6.9)$$

where the Fourier transform of G_t is given by

$$\hat{G}_t(\lambda) = \sum_{k=0}^{\infty} e^{-(2k+n)|\lambda|t} m((2k+n)|\lambda|) P_k(\lambda).$$

Differentiating (2.6.9) k times with respect to t and once with respect to s, and putting $s = t$, we get

$$\partial_s^{k+1} T^{2s}(Mf) = F_s * \partial_s T^s f$$

where F_s is given by

$$\hat{F}_s(\lambda) =$$

$$(-1)^k \sum_{k=0}^{\infty} ((2k+n)|\lambda|)^k \, e^{-(2k+n)|\lambda|s} m((2k+n)|\lambda|) P_k(\lambda).$$

Therefore, we have the inequality

$$|\partial_s^{k+1} T^{2s}(Mf)(w)| \leq \int |F_s(v)| |\partial_s T^s f(v^{-1}w)| \, dv.$$

Applying Cauchy-Schwarz inequality we obtain

$$|\partial_s^{k+1} T^{2s}(Mf)(w)|^2$$

$$\leq A(s) \left(\int_{H^n} \left(1 + s^{-2}|v|^4\right)^{-k} |\partial_s T^s f(v^{-1}w)|^2 \, dv \right)$$

where we have written

$$A(s) = \int_{H^n} \left(1 + s^{-2}|v|^4\right)^k |F_s(v)|^2 \, dv.$$

Now we assume that a slightly stronger condition is satisfied by the function m namely, the condition on its derivatives hold for all $j \leq n + 2$ when n is even and $j \leq n + 3$ when n is odd. Under these conditions on m we claim that

$$A(s) \leq Cs^{-n-2k-1}.$$

Assuming the claim for a moment, we have

$$|\partial_s^{k+1} T^{2s}(Mf)(w)|^2$$

$$\leq Cs^{-n-2k-1} \int_{H^n} \left(1 + s^{-2}|v|^4\right)^{-k} |\partial_s T^s f(v^{-1}w)|^2 \, dv$$

and integrating this against $s^{2k+1} ds$, we get the pointwise estimate

$$g_{k+1}(Mf, w) \leq C g_k^*(f, w)$$

and this will complete the proof of the multiplier theorem.

Coming to the proof of the claim, let us write

$$A_1(s) = \int_{|w| \leq \sqrt{s}} \left(1 + s^{-2}|w|^4\right)^k |F_s(w)|^2 \, dw$$

and let $A_2(s)$ be the integral taken over $|w| \geq \sqrt{s}$. Estimating $A_1(s)$ is easy: as $|w| \leq \sqrt{s}$ we have

$$A_1(s) \leq C \int_{H^n} |F_s(w)|^2 \, dw$$

and hence by the Plancherel theorem, $A_1(s)$ is less than or equal to

$$\leq C \int_{-\infty}^{\infty} \left(\sum_{j=0}^{\infty} (2j+n)^{2k} |\lambda|^{2k} e^{-2s(2j+n)|\lambda|} \frac{(k+n-1)!}{j!} \right) d\mu(\lambda)$$

which is dominated by

$$Cs^{-n-2k-1} \sum_{j=0}^{\infty} (2j+n)^{-2}.$$

This proves our claim on the part of the integral taken over $|w| \leq \sqrt{s}$.
Now consider $A_2(s)$. With $w = (z, s)$ we observe that

$$A_2(s) \leq Cs^{-2k} \int_{H^n} (t^2 + |z|^4)^k |F_s(z, t)|^k \, dz \, dt$$

which is dominated by the integral

$$s^{-2k} \int_{H^n} |(it - \frac{1}{4}|z|^2)^k F_s(z, t)|^2 \, dz \, dt.$$

As $F_s(z, t)$ is radial, Plancherel theorem applied to F_s shows that the above integral equals a constant times

$$\int_{-\infty}^{\infty} \left(\sum_{j=0}^{\infty} |R_j(\lambda, (it - \frac{1}{4}r^2)^k F_s(r, t)|^2 \frac{(j + n - 1)!}{j!} \right) d\mu(\lambda)$$

and so we will be done if we could show that the above is dominated by a constant times s^{-n-1}.

As we did in the case of the Bochner-Riesz kernel, we express

$$R_j(\lambda, (it - \frac{1}{4}r^2)^k F_s(r, t))$$

in terms of $R_j(\lambda, F_s(r, t))$. To do this let us set

$$\psi(j, \lambda) = (-1)^k (2j + n)^k |\lambda|^k e^{-s(2j+n)|\lambda|} m((2j + n)^k |\lambda|)$$

so that $R_j(\lambda, F_s(r, t)) = \psi(j, \lambda)$. We define

$$\psi_m(j, \lambda) = R_j(\lambda, (it - \frac{1}{4}r^2)^m F_s(r, t)).$$

Then the estimate for $A_2(s)$ follows from the next lemma.

Lemma 2.6.4 *Under the assumptions we have on* m,

$$|\psi_k(j, \lambda)| \leq Ce^{-\epsilon(2j+n)|\lambda|s}$$

for some $\epsilon > 0$.

Proof: Consider

$$\psi_1(j, \lambda) = R_1(\lambda, (it - \frac{1}{4}r^2) F_s(r, t)).$$

Assuming $\lambda > 0$ and proceeding as in the previous section, we get

$$\psi_1(j,\lambda) = \frac{\partial \psi}{\partial \lambda} - \frac{j}{\lambda}\left(\psi(j,\lambda) - \psi(j-1,\lambda)\right).$$

Since $\psi(j,\lambda) = \psi((2j+n)|\lambda|)$ we can write the above in the form

$$\psi_1(j,\lambda) = \frac{n}{2\lambda}\frac{\partial \psi}{\partial j} + \frac{j}{\lambda}\left(\frac{\partial \psi}{\partial j} - \Delta_-\psi\right)$$

where

$$\Delta_-\psi(j,\lambda) = \psi(j,\lambda) - \psi(j-1,\lambda).$$

Define operators S, D and T by

$$S\psi = \frac{\partial \psi}{\partial j}, \quad D\psi = \frac{\partial \psi}{\partial j} - \Delta_-\psi, \quad T\psi = jD\psi$$

so that we have

$$\psi_1(j,\lambda) = \lambda^{-1}\left(\frac{n}{2}S + T\right)\psi(j,\lambda).$$

Iteration of this formula gives us

$$\psi_k(j,\lambda) = \lambda^{-k} \sum_{i+l+q=k} C_{ilq}S^i T^l S^q \psi(j,\lambda).$$

We now observe that

$$S^q\psi(j,\lambda) = \psi^{(q)}((2j+n)\lambda)(2\lambda)^q$$

and by the hypothesis on m, the operator S^q in essence brings down a factor $(2j+n)^{-q}$. We will show that T does the same. Then each term in the sum will behave like

$$\lambda^{-k}(2j+n)^{-k}\psi(j,\lambda).$$

Recalling the definition of $\psi(j,\lambda)$, we conclude that

$$|\psi_k(j,\lambda)| \leq Ce^{-\epsilon(2j+n)|\lambda|s}$$

as desired. For the operators T^l, the following formula is valid.

Lemma 2.6.5

$$T^l\psi(j,\lambda) = \sum C_{pqs}j^p D^q \Delta_-^s \psi(j,\lambda)$$

the sum being extended over all p, q, s with $l + p \leq 2q + s \leq 2l$.

Proof: We prove this lemma by induction. We first observe that from the definition of T the lemma is trivially true for $l = 1$. Now assume the lemma for some l and consider $T^{l+1}\psi$.

$$T^{l+1}\psi(j,\lambda) = \sum C_{pqs} j \left(j^p D^q \Delta_-^s \psi(j,\lambda) \right)$$

where $l + p \leq 2q + s \leq 2l$. We claim that

$$D(j^p D\psi(j,\lambda)) =$$

$$j^p D^2 \psi(j,\lambda) + \sum_{r=0}^{p-1} a_r j^r D\Delta_\psi(j,\lambda) + \sum_{r=0}^{p-2} b_r j^r D\psi(j,\lambda).$$

Assuming the claim for a moment we see that

$$T^{l+1}\psi = \sum C_{pqs} j^{p+1} D^{q+1} \Delta_-^s \psi(j,\lambda) +$$

$$\sum C_{pqs} \left(\sum_{r=0}^{p-1} a_r j^{r+1} D^q \Delta_-^{s+1} \psi(j,\lambda) + \sum_{r=0}^{p-2} b_r j^{r+1} D^q \Delta_-^s \psi(j,\lambda) \right).$$

From this it is easy to see that T^{l+1} has the desired form. To prove the claim we first observe that

$$\Delta_-(\varphi\psi)(j) = \Delta_-\varphi(j)\psi(j) + \varphi(j-1)\Delta_-\psi(j)$$

and from this formula we get

$$\Delta_-(j^p D\psi) = \Delta_-(j^p)D\psi + (j-1)^p D(\Delta_-\psi).$$

We also have the equations

$$\Delta_-(j^p) = pj^{p-1} - \sum_{r=0}^{p-2} b_r j^r,$$

$$(j-1)^p D(\Delta_-\psi) = j^p D(\Delta_-\psi) - \sum_{r=0}^{p-1} a_r j^r D(\Delta_-\psi),$$

and

$$\frac{\partial}{\partial j}(j^p D\psi) = pj^{p-1} D\psi + j^p D(\frac{\partial}{\partial j}\psi).$$

Putting these things together we see that our claim is true.

Finally, it still remains to prove that the action of T^l has the desired properties. Using Taylor's formula with integral form of remainder, we have

$$D\psi(j, \lambda) = \int_0^1 t\psi''(j - 1 + t, \lambda) \, dt$$

where the primes stand for derivatives with respect to j. From the above, it is clear that each time we apply D, we bring down a factor of j^{-2}. An iteration shows that D^q will bring down j^{-2q} and as Δ_-^s brings down j^{-s}, the above lemma shows that T^l acting on ψ will in effect bring down

$$\sum_{pqs} C_{pqs} j^p j^{-2q-s}.$$

Since $p + l \le 2q + s \le 2l$, T^l brings down j^{-l} as required. ∎

2.7 Notes and references

In two long papers [70], [71] Strichartz developed the harmonic analysis on H^n as the joint spectral theory of the sublaplacian and T. We have closely followed him in Sections 1 and 2. A good source of information on the restriction theorem is the book [65] of Stein. For versions of the restriction theorem for spherical harmonic expansions and more generally for eigenfunction expansions of the Laplace-Beltrami operators on compact Riemannian manifolds, we refer to Sogge [62]. Applications of the restriction theorem to the study of Bochner-Riesz means and to some problems in partial differential equations can also be found in [62]. Theorem 2.3.2 is due to Müller [44]. Some extensions were proved by the author in [80] and [81]. The estimates for the special Hermite projections were taken from Ratnakumar et al [54]. Theorem 4.1 is in [85] but there the converse was proved only for radial functions. The Bochner-Riesz means for the sublaplacian was first studied by Mauceri [39]. Later Müller [45] and the author [78] studied them using different techniques. General multipliers for the Fourier transform were studied by de Michele and Mauceri [41]. Multipliers of the form $m(H(\lambda))$ have been studied by Mauceri [42], Christ [13] and others; the optimal result was proved recently by Müller and Stein [46]. The proof presented here is taken from [79]. For more about the sublaplacian and analysis on nilpotent groups, we refer to Folland [25], Folland-Stein [29], Taylor [74] and Beals-Greiner [5].

Chapter 3

GROUP ALGEBRAS AND APPLICATIONS

Even though the algebra $L^1(H^n)$ is not commutative, the subalgebra $L^1(H^n/U(n))$ of radial functions forms a commutative Banach algebra under convolution. In this chapter we study the Gelfand transform on this algebra. The Gelfand spectrum is identified with the set of all bounded $U(n)$-spherical functions which are given by Bessel and Laguerre functions. We also consider the Banach algebra generated by the surface measures μ_r and get optimal estimates for its characters, from which we proceed to study Wiener-Tauberian theorems and spherical means. We prove a one radius theorem for the spherical means using the summability result of Strichartz proved in the previous chapter. We also prove a maximal theorem for the spherical means on the Heisenberg group.

3.1   The Heisenberg motion group

In the previous chapter we obtained a decomposition of functions in terms of the eigenfunctions of the sublaplacian from the spectral theory perspective. There is also a representation theoretic interpretation of the spectral decomposition. For this we need a larger group, G^n, called the Heisenberg motion group. This terminology is justified by the fact that this is the group of isometries for the natural Heisenberg geometry for which $\mathcal{L}$ is the Laplacian. (See the works of Koranyi [36] and Strichartz [70].) This group is the semidirect product of $U(n)$ and H^n. Thus $G^n = U(n) \times H^n$ as a set. We know that for each $\sigma \in U(n)$ the map $(z,t) \to (\sigma z, t)$ is an automorphism of the Heisenberg group. The Heisenberg motion group acts on H^n by

$$\pi(\sigma, z, t)(w, s) = (z, t)(\sigma w, s)$$

and the group law of G^n is given by

$$(\sigma, z, t)(\tau, w, s) = (\sigma\tau, z + \sigma w, t + s - \frac{1}{2}Im\sigma w.\bar{z}).$$

The Heisenberg group H^n and the unitary group $U(n)$ are subgroups of the Heisenberg motion group G^n. Moreover, $U(n)$ is a normal subgroup of G^n and H^n can be identified with the quotient group $G^n/U(n)$.

Since G^n commutes with $\mathcal{L}$ and T, it acts on the joint eigenfunctions of these operators. Thus to each point $(\lambda, (2k + n)|\lambda|)$ on the Heisenberg fan, there is an associated representation ρ_k^λ of G^n. We will show that these representations are irreducible. In fact, when restricted to H^n, the representations ρ_k^λ are primary, consisting of a finite number of copies of the Schrödinger representation associated to the parameter $-\lambda$. So our spectral theory can be reinterpreted as a decomposition of $L^2(H^n)$ into irreducible representations of G^n under the induced action of G^n on $L^2(H^n)$.

The representation ρ_k^λ associated to the point $(\lambda, (2k + n)|\lambda|)$ on the Heisenberg fan is realised on a Hilbert space of eigenfunctions for the sublaplacian. Indeed, let

$$E_{\alpha,\beta}^\lambda(z, t) = (\pi_\lambda(z, t)\Phi_\alpha^\lambda, \Phi_\beta^\lambda)$$

be the matrix components associated to the Schrödinger representation π_λ. Let $\mathcal{H}_k^\lambda$ be the Hilbert space for which

$$\{E_{\alpha,\beta}^\lambda : \alpha, \beta \in N^n, |\alpha| = k\}$$

is an orthonormal basis. The elements of this space $\mathcal{H}_k^\lambda$ can be characterised by the eigenvalue equations

$$\mathcal{L}f = (2k + n)|\lambda|f, \quad -iTf = \lambda f$$

and the condition $f(z, 0) \in L^2(\mathbb{C}^n)$. We define an inner product on $\mathcal{H}_k^\lambda$ by setting

$$(f, g) = (2\pi)^{-n}|\lambda|^n \int_{\mathbb{C}^n} f(z, 0)\bar{g}(z, 0)\, dz.$$

The action of ρ_k^λ on $\mathcal{H}_k^\lambda$ is given by

$$\rho_k^\lambda(\sigma, z, t)f(w, s) = f(\pi((\sigma, z, t)^{-1})(w, s)).$$

The properties of these representations are summarised in the following theorem.

Theorem 3.1.1 *The representation ρ_k^λ defined above is an irreducible unitary representation of G^n. The restriction of ρ_k^λ to H^n breaks up into a sum of $\frac{(k+n-1)!}{k!(n-1)!}$ irreducible representations each of which is equivalent to the representation $\pi_{-\lambda}$ of H^n. Moreover, for λ, $\lambda' \in \mathbb{R}^*$, $k, k' \in N$, ρ_k^λ is equivalent to $\rho_{k'}^{\lambda'}$ if and only if $k = k'$ and $\lambda = \lambda'$.*

Proof: It is clear that ρ_k^λ is a unitary representation. To show that it is irreducible, we assume that $\lambda = 1$. We treat the general case in a similar fashion. Let M_0 be a closed subspace of $\mathcal{H}_k^1$ which is invariant under $\rho_k^1(\sigma, z, t)$ for all $(\sigma, z, t) \in G^n$. We will show that $M_0 = \mathcal{H}_k^1$, proving the irreducibility of ρ_k^1. As

$$\{E_{\alpha,\beta}^1 : \alpha, \beta \in N^n, |\alpha| = k\}$$

is an orthonormal basis for $\mathcal{H}_k^1$, any function in M_0 is of the form $e^{is} f(w)$ with

$$f(w) = \sum_{|\alpha|=k} \sum_\beta c_{\alpha,\beta} \Phi_{\alpha,\beta}(w)$$

belonging to $L^2(\mathbb{C}^n)$. Let

$$M = \{f(w) : e^{is} \bar{f}(w) \in M_0\}.$$

Then M is a closed subspace of $L^2(\mathbb{C}^n)$ and $f \times \varphi_k = f$ for any $f \in M$ since $\bar{f}$ is in the span of $\Phi_{\alpha,\beta}$ with $|\alpha| = k$. Moreover,

$$\rho_k^1(\sigma, z, t) M_0 \subset M_0$$

means $\rho_k^1((\sigma, z, t)^{-1}) F(w, s) \in M_0$ whenever $F(w, s) \in M_0$ for all $(\sigma, z, t) \in G^n$. If $F(w, s) = e^{is} f(z)$ then the above condition translates into

$$e^{i(s+t)} e^{\frac{i}{2} Im(z.\overline{\sigma w})} f(z + \sigma w) \in M_0.$$

From this we infer that the closed subspace M is invariant under rotations and twisted translations, i.e.,

$$e^{-\frac{i}{2} Im(z.\bar{w})} f(z + w) \in M$$

for all $z \in \mathbb{C}^n$ whenever $f \in M$.

We first claim that $\varphi_k \in M$. If not we can find $g \in L^2(\mathbb{C}^n)$ such that $(f, g) = 0$ for all $f \in M$ and $(\varphi_k, g) \neq 0$. As φ_k is radial and M is rotation invariant, we can assume that g is radial. Since M is invariant under the twisted translations and g is radial, it follows that $f \times \bar{g} = 0$ for all $f \in M$. But then

$$0 = f \times \bar{g} \times \varphi_k = (2\pi)^{-n} \frac{k!(n-1)!}{(k+n-1)!} (\varphi_k, \bar{g}) f \times \varphi_k.$$

But this is not possible as $(\varphi_k, \bar{g}) \neq 0$ and $f \times \varphi_k = f$ for all $f \in M$. This proves our claim.

Next we claim $\overline{\Phi_{\alpha,\beta}} \in M$ for all β, α with $|\alpha| = k$. To see this let V be the orthogonal complement of M and $g \in V$. As M is rotation invariant the function $g(-w) \in V$ as well. Now $\varphi_k \in M$ and M is invariant under the twisted translations

$$f(w) \to e^{-\frac{i}{2} Im(z.\bar{w})} f(z + w).$$

Therefore, it follows that $\varphi_k \times \bar{g} = 0$ which is the same as $g \times \varphi_k = 0$. Taking the Weyl transform, we get $W(g)P_k = 0$. Therefore, for any α with $|\alpha| = k$ and $\beta \in N^n$, we have

$$(W(g)P_k\Phi_\alpha, \Phi_\beta) = 0$$

which simply means that

$$\int_{\mathbb{C}^n} g(w)\Phi_{\alpha,\beta}(w) \, dw = 0$$

for all g in the orthogonal complement of M. This proves that $\overline{\Phi_{\alpha,\beta}} \in M$ whenever $|\alpha| = k$. Thus M_0 has to be the whole space $\mathcal{H}_k^1$, and this proves the irreducibility of ρ_k^1.

Let $e \in U(n)$ be the identity matrix. The restriction of ρ_k^λ to H^n is given by

$$\rho_k^\lambda(e, z, t)F(w, s) = e^{i\lambda(s - t - \frac{1}{2} Im(z.\bar{w}))} f(w - z)$$

where $F(w, s) = e^{i\lambda s} f(w)$. For every α with $|\alpha| = k$, let $\mathcal{H}_\alpha^\lambda$ be the subspace of $\mathcal{H}_k^\lambda$ containing functions of the form

$$F(w, s) = \sum_\beta c_\beta(\pi_\lambda(w, s)\Phi_\alpha, \Phi_\beta).$$

Define an operator

$$U : L^2(\mathbb{R}^n) \to \mathcal{H}_\alpha^\lambda$$

by the prescription $U\Phi_\beta^\lambda = E_{\alpha,\beta}^\lambda$. As U takes one orthonormal basis into another, it is clearly unitary. We claim that

$$U\pi_\lambda(-\bar{z}, -t)U^* = \rho_k^\lambda(e, z, t).$$

Since $(z, t) \to (-\bar{z}, -t)$ is an automorphism of the Heisenberg group

$$(z, t) \to \pi_\lambda(-\bar{z}, -t)$$

is a representation of H^n which is unitarily equivalent to $\pi_{-\lambda}(z, t)$. This shows that the restriction of ρ_k^λ to H^n is unitarily equivalent to $\pi_{-\lambda}$.

To prove the claim it is enough to show that

$$U\pi_\lambda(-\bar{z}, -t)U^*E_{\alpha,\beta}^\lambda = \rho_k^\lambda(e, z, t)E_{\alpha,\beta}^\lambda$$

for all β. Again, let us assume $\lambda = 1$. As $U^*E_{\alpha,\beta}^1 = \Phi_\beta$ we have

$$\pi_1(z, t)U^*E_{\alpha,\beta}^1(\xi) = e^{it}e^{i(x.\xi + \frac{1}{2}x.y)}\Phi_\beta(\xi + y).$$

Expanding this function in terms of Φ_μ, we get

$$\pi_1(z, t)U^*E_{\alpha,\beta}^1(\xi) = (2\pi)^{\frac{n}{2}}e^{it}\sum_\mu \Phi_{\beta,\mu}(z)\Phi_\mu(\xi).$$

Applying U and recalling its action, we get

$$U\pi_1(-\bar{z}, -t)U^*E_{\alpha,\beta}^1(w, s) =$$

$$= (2\pi)^n e^{-it}e^{is}\sum_\mu \overline{\Phi_{\beta,\mu}}(z)\Phi_{\alpha,\mu}(w). \qquad (3.1.1)$$

Here we have used the fact that $\overline{\Phi_{\beta,\mu}}(z) = \Phi_{\beta,\mu}(-\bar{z})$ which follows from the definition.

On the other hand, let us look at

$$\rho_k^1(e, z, t)E_{\alpha,\beta}^1(w, s)$$

$$= (2\pi)^{\frac{n}{2}}e^{i(s-t-\frac{1}{2}Im(z.\bar{w}))}\Phi_{\alpha,\beta}(w - z).$$

We can expand the function

$$F_z(w) = e^{-\frac{i}{2}Im(z.\bar{w})}\Phi_{\alpha,\beta}(w - z)$$

in terms of special Hermite functions. Let us calculate

$$(F_z, \Phi_{\mu,\nu}) = \int_{\mathbb{C}^n} e^{-\frac{i}{2} Im(z.\bar{w})} \Phi_{\alpha,\beta}(w - z) \bar{\Phi}_{\mu,\nu}(w) \, dw.$$

Making a change of variable and noting that $\bar{\Phi}_{\beta,\alpha}(z - w) = \Phi_{\alpha,\beta}(w - z)$, we get

$$\overline{(F_z, \Phi_{\mu,\nu})} = \Phi_{\beta,\alpha} \times \Phi_{\mu,\nu}(z)$$

which in view of the orthogonality relations between special Hermite functions gives us the formula

$$(F_z, \Phi_{\mu,\nu}) = (2\pi)^{\frac{n}{2}} \delta_{\alpha,\mu} \overline{\Phi_{\beta,\nu}}(z).$$

Thus we have proved

$$\rho_k^1(e, z, t) E_{\alpha,\beta}^1(w, s)$$
$$= (2\pi)^{\frac{n}{2}} e^{-it} e^{is} \sum_\mu \overline{\Phi_{\beta,\mu}}(z) \Phi_{\alpha,\mu}(w).$$

Comparing this with (3.1.1) we conclude that

$$U\pi_1(-\bar{z}, -t) U^* = \rho_k^1(e, z, t).$$

Since π_λ is irreducible, so is ρ_k^λ restricted to $\mathcal{H}_\alpha^\lambda$. The space $\mathcal{H}_k^\lambda$ is the direct sum of $\mathcal{H}_\alpha^\lambda$, $|\alpha| = k$ and there are exactly $\frac{(k+n-1)!}{k!(n-1)!}$ multi-indices α with $|\alpha| = k$. This proves that the restriction of ρ_k^λ to H^n breaks up into a sum of $\frac{(k+n-1)!}{k!(n-1)!}$ irreducible representations each of which is unitarily equivalent to $\pi_{-\lambda}$. Finally, considering their restrictions to H^n, it follows that ρ_k^λ and $\rho_k^{\lambda'}$ are inequivalent when $\lambda \neq \lambda'$. Similarly by considering their restrictions to $U(n)$ we can conclude that when $k \neq m$ the representations ρ_k^λ and ρ_m^λ are inequivalent. This completes the proof of the theorem. ∎

The action of ρ_k^λ when restricted to $U(n)$ is given by

$$\rho_k^\lambda(\sigma, 0, 0) F(w, s) = F(\sigma^{-1} w, s).$$

Thus the class of radial functions in $\mathcal{H}_k^\lambda$ are left invariant by $\rho_k^\lambda(\sigma, 0, 0)$. This means that ρ_k^λ is a class-1 representation for the pair $(G^n, U(n))$. In $\mathcal{H}_k^\lambda$ there is exactly one radial function (up to constant multiples) which is given by

$$e_k^\lambda(w, s) = e^{i\lambda s} \varphi_k^\lambda(w).$$

We now consider the Fourier transform on G^n and prove a restriction theorem for the Heisenberg motion group.

Given a function F on G^n, we can define the Fourier transform of F in the usual way. For each representation ρ_k^λ, we define

$$\rho_k^\lambda(F) = \int_{G^n} F(g)\rho_k^\lambda(g)\, dg$$

as an operator on the Hilbert space $\mathcal{H}_k^\lambda$. Any function $f(z, t)$ on H^n can be considered as a right $U(n)$-invariant function on G^n. For such functions it is easy to calculate $\rho_k^\lambda(f)$. Let $\varphi(w, s) = e^{i\lambda s}\varphi(w)$. It follows from the definition of these representations that

$$\rho_k^\lambda(f)\varphi(w, s) = e^{i\lambda s} f^{-\lambda} *_\lambda \Phi(w)$$

where Φ is the radialisation of φ given by

$$\Phi(w) = \int_{U(n)} \varphi(\sigma w)\, d\sigma.$$

Using this we can easily calculate the Hilbert-Schmidt operator norm of $\rho_k^\lambda(f)$.

Consider the function

$$\varphi(w)e^{i\lambda s} = (\pi_\lambda(w, s)\Phi_\alpha^\lambda, \Phi_\beta^\lambda) = E_{\alpha,\beta}^\lambda(w, s)$$

in $\mathcal{H}_k^\lambda$. Since e_k^λ is the unique radial function in $\mathcal{H}_k^\lambda$ we see that

$$\Phi(w)e^{i\lambda s} = Ce_k^\lambda(w, s)$$

for some constant C. When $\alpha \neq \beta$, $(\pi_\lambda(0, s)\Phi_\alpha^\lambda, \Phi_\beta^\lambda) = 0$ which means that $\Phi(w) = 0$ for $\alpha \neq \beta$. When $\alpha = \beta$ we easily see that

$$\Phi(w)e^{i\lambda s} = (2\pi)^{\frac{n}{2}} \frac{k!(n-1)!}{(k+n-1)!} e_k^\lambda(w, s).$$

Therefore, the Hilbert-Schmidt operator norm of $\rho_k^\lambda(f)$ is given by

$$\|\rho_k^\lambda(f)\|_{HS}^2 = \sum_{|\alpha|=k} \|\rho_k^\lambda(f)E_{\alpha,\alpha}^\lambda\|_{\mathcal{H}_k^\lambda}^2$$

which in view of the above calculation gives us the following result.

Proposition 3.1.2 *For* $f \in L^1 \cap L^2(H^n)$, $\rho_k^\lambda(f)$ *is a Hilbert-Schmidt operator on* $\mathcal{H}_k^\lambda$ *and*

$$\|\rho_k^\lambda(f)\|_{HS}^2 = |\lambda|^n \frac{k!(n-1)!}{(k+n-1)!} \int_{C^n} |f^\lambda *_\lambda \varphi_k^\lambda(w)|^2 \, dw.$$

From this proposition we immediately see that the Plancherel theorem can be written in the form

$$\|f\|_2^2 = (2\pi)^{-2n-1}$$

$$\times \sum_{k=0}^\infty \left(\int_{-\infty}^\infty \|\rho_k^\lambda(f)\|_{HS}^2 |\lambda|^n \, d\lambda \right) \frac{(k+n-1)!}{k!(n-1)!}.$$

This is the representation theoretic interpretation of Theorem 1.1.2. We conclude this section with the following restriction theorem. Recall the definition of $\mathcal{S}_q$ from Chapter 1.

Theorem 3.1.3 *Let* $0 < \gamma < \frac{3n-2}{3n+4}$ *and let* $1 \leq p \leq 1 + \gamma$. *Then*

$$\|\rho_k^\lambda(f)\|_{\mathcal{S}_q} \leq C_\lambda k^{-\frac{(3n-2)n}{(3n+1)q}} \|f\|_{(p,1)}$$

for all $f \in L^{(p,1)}(H^n)$ *where* $q = \frac{2\gamma}{1+\gamma} p'$.

Proof: The proof is similar to the proof of Theorem 1.3.3 . From the definition it follows that

$$\|\rho_k^\lambda(f)\|_{\mathcal{S}_\infty} \leq \|f\|_{(1,1)}.$$

And the proposition gives

$$\|\rho_k^\lambda(f)\|_{HS}^2 \leq C_\lambda k^{-n+1} \|f^\lambda *_\lambda \varphi_k^\lambda\|_2^2.$$

Expressing the λ-twisted convolution in terms of the twisted convolution and using the estimates of Proposition 2.3.4, we get

$$\|\rho_k^\lambda(f)\|_{HS} \leq C_\lambda k^{\frac{n}{p}-n} \|f\|_{(p,1)}$$

for $1 \leq p \leq 1 + \gamma$. Appealing to the noncommutative interpolation theorem of Peetre-Sparr [52], we conclude the proof. ∎

3.2 Gelfand pairs, spherical functions and group algebras

Suppose we are given a Lie group G and a compact Lie subgroup K of $Aut(G)$, the group of automorphisms of G. There is a natural action of K on the convolution algebra $L^1(G)$. Let us denote by $L^1(G/K)$ the subalgebra of those elements of $L^1(G)$ that are invariant under the action of K. The pair (G, K) is said to be a Gelfand pair if the algebra $L^1(G/K)$ is commutative.

In the traditional definition of a Gelfand pair, one assumes that K is a compact subgroup of G. One then defines (G, K) to be a Gelfand pair if the set of K-bi-invariant functions in $L^1(G)$ forms a commutative Banach algebra. That our definition is equivalent to the traditional one can be easily seen by considering K as a compact subgroup of the semidirect product $K \times G$. Then the K-bi-invariant functions on $K \times G$ are precisely the functions on G that are invariant under the action of K.

Associated to each Gelfand pair are certain functions called $K-$ spherical functions. There are several equivalent ways of defining $K-$ spherical functions associated to a Gelfand pair. A $K-$ spherical function is a smooth complex valued $K-$ invariant function φ on G such that $\varphi(e) = 1$ and φ is a joint eigenfunction of all left invariant $K-$ invariant differential operators on G. Equivalently, they can be defined as complex homomorphisms of the commutative Banach algebra $L^1(G/K)$. They are characterised by the property

$$\int_K \varphi(a\,k.b)\,dk = \varphi(a)\varphi(b) \qquad (3.2.1)$$

for all $a, b \in G$. In the above dk is the Haar measure on K.

In this section we study K-spherical functions on the Heisenberg group. The unitary group $U(n)$ gives rise to a subgroup of the automorphism group $Aut(H^n)$ by the action $\sigma(z,t) = (\sigma z, t)$ for $\sigma \in U(n)$. We denote this subgroup by the same symbol, $U(n)$. It is a compact subgroup of $Aut(H^n)$. Conjugating by an automorphism if necessary, we can always assume that a given connected compact subgroup of $Aut(H^n)$ is contained in $U(n)$. There are many subgroups K of $U(n)$ for which (H^n, K) is a Gelfand pair. As we have remarked earlier, (H^n, K) will be a Gelfand pair if and only if $(K \times H^n, K)$ is a Gelfand pair in the traditional sense. When $K = U(n)$ we are led to the Heisenberg motion group G^n of the previous section.

There is a general theory of K-spherical functions on the Heisenberg group developed by Benson et al [7]. As we are mainly interested in the cases when $K = U(n)$ or $K = T(n)$, we do not need the full force of this theory. However, we recall without proof a result from their theory which is useful in parametrising the K-spherical functions. Given a strongly continuous unitary representation π of G, let K_π be the subgroup of all $\sigma \in K$ for which the representation $\pi_\sigma(g) : g \to \pi(\sigma g)$ is unitarily equivalent to π. For $\sigma \in K_\pi$ we choose an intertwining operator $m_\pi(\sigma)$ such that

$$\pi(\sigma g) = m_\pi(\sigma)\pi(g)m_\pi(\sigma)^*.$$

Let $\mathcal{H}_\pi$ be the representation space for π and let $\mathcal{H}_\pi = \sum \mathcal{P}_\alpha$ be the decomposition of $\mathcal{H}_\pi$ into irreducible subspaces invariant under the action of m_π. With these notations we have the following result.

Theorem 3.2.1 *If φ is a bounded K-spherical function, then it is of the form*

$$\varphi(g) = \varphi_{\pi,v}(g) = \int_K (\pi(\sigma g)v, v)\, d\sigma$$

for some irreducible unitary representation π and a unit vector $v \in \mathcal{P}_\alpha$. Moreover,

$$\varphi_{\pi,v}(g) = \varphi_{\pi',v'}(g)$$

if and only if π is unitarily equivalent to π'_σ for some $\sigma \in K$ and v, v' belong to the same $\mathcal{P}_\alpha$.

Corollary 3.2.2 *Suppose that $K_\pi = K$ and $\{v_1, v_2, \ldots, v_l\}$ is an orthonormal basis for $\mathcal{P}_\alpha$. Then*

$$\varphi_{\pi,\alpha}(g) = \frac{1}{l}\sum_{j=1}^{l}(\pi(g)v_j, v_j)$$

where $\varphi_{\pi,\alpha}(g) = \varphi_{\pi,v}(g)$ with $v \in \mathcal{P}_\alpha$ and $\|v\| = 1$.

We use the above theorem and the corollary to find all the K-spherical functions when $K = U(n)$ or $T(n)$. Here

$$T(n) = \{e^{i\theta} = (e^{i\theta_1}, e^{i\theta_2}, \ldots, e^{i\theta_n}) : \theta_j \in \mathbb{R}\}$$

is the n-torus which acts by

$$e^{i\theta}z = (e^{i\theta_1}z_1, e^{i\theta_2}z_2, \ldots, e^{i\theta_n}z_n).$$

We first describe all the K-spherical functions associated to the infinite dimensional representations π_λ.

Proposition 3.2.3 *The bounded $T(n)$-spherical functions coming from π_λ are given by*

$$E_\alpha^\lambda = (2\pi)^{\frac{n}{2}} e^{i\lambda t} \Phi_{\alpha,\alpha}(\sqrt{|\lambda|}z), \quad \alpha \in N^n.$$

Proof: For $\sigma \in U(n)$ consider the representation

$$\pi_{\lambda,\sigma}(z,t) = \pi_\lambda(\sigma z, t).$$

As they agree at the centre, they are unitarily equivalent. So we can choose an intertwining operator $\mu_\lambda(\sigma)$ which, when $\lambda = 1$, was called the metaplectic representation. We can appeal to the above corollary to compute the spherical functions. When $e^{i\theta} \in T(n)$ we know that $\mu_\lambda(e^{i\theta})$ is diagonalised by the Hermite basis

$$\{\Phi_\alpha^\lambda : \alpha \in N^n\}.$$

Thus for each multi-index α the space $\mathcal{P}_\alpha$ is spanned by the single Φ_α^λ. Therefore,

$$\varphi_{\lambda,\alpha}(z,t) = (\pi_\lambda(z,t)\Phi_\alpha^\lambda, \Phi_\alpha^\lambda)$$

which is nothing but $(2\pi)^{\frac{n}{2}} e^{i\lambda t} \Phi_{\alpha,\alpha}(\sqrt{|\lambda|}z)$. This proves the proposition.
∎

Thus the spherical functions for the Gelfand pair $(H^n, T(n))$ are given by the special Hermite functions. Later in this section we will actually verify that these functions have the characterising property of spherical functions. In the next proposition we identify all $U(n)$ spherical functions.

Proposition 3.2.4 *The bounded $U(n)$-spherical functions coming from π_λ are given by*

$$E_k^\lambda(z,t) = \frac{k!(n-1)!}{(k+n-1)!} e^{i\lambda t} \varphi_k^\lambda(z), \quad k \in N.$$

Proof: Again we work with the metaplectic representation μ_λ. For $\sigma \in U(n)$ we use the fact that the k-th eigenspace spanned by

$$\{\Phi_\alpha^\lambda : |\alpha| = k\}$$

is invariant under $\mu_\lambda(\sigma)$. This means that the spherical functions are indexed by $k \in N$ and are given by

$$\varphi_{\lambda,k}(z,t) = d_k^{-1} \sum_{|\alpha|=k} (\pi_\lambda(z,t)\Phi_\alpha^\lambda, \Phi_\alpha^\lambda)$$

where d_k is the dimension of the k-th eigenspace. Since

$$\sum_{|\alpha|=k} \Phi_{\alpha,\alpha}(z) = (2\pi)^{-\frac{n}{2}} \varphi_k(z)$$

and $d_k^{-1} = \frac{k!(n-1)!}{(k+n-1)!}$ the proposition follows.   ∎

We now consider the one-dimensional representations χ_w parametrised by $w \in \mathbb{C}^n$. Recall that $\chi_w(z,t)$ acts on $\mathbb{C}$ by multiplication by the complex number $e^{iRe(w.\bar{z})}$. If (H^n, K) is a Gelfand pair, then for every χ_w, we have a K-spherical function

$$\varphi_w(z,t) = \int_K e^{iRe(w.\overline{k.z})}\, dk.$$

Thus the spherical function $\varphi_w(z,t)$ is independent of t and depends only on the K-orbit through w. By Theorem 3.2.1, $\varphi_w \neq \varphi_{w'}$ if w' does not belong to the K-orbit of w. Let $\mu_{K.w}$ denote the unit measure supported on the K-orbit through w. As a distribution, this is given by

$$(\mu_{K.w}, f) = \int_K f(k^{-1}.w, 0)\, dk$$

for all $f \in C_0^\infty(H^n)$. The Euclidean Fourier transform of f can be written in the form

$$\hat{f}(w,s) = (2\pi)^{-\frac{2n+1}{2}}$$

$$\times \int_{H^n} f(z,t) e^{-i(Re(w.\bar{z})+ts)}\, dzdt.$$

Now

$$\int_{H^n} \varphi_w(z,t) f(z,t)\, dzdt =$$

$$\int_{H^n} \int_K f(z,t) e^{iRe(w.\overline{k.z})}\, dkdzdt$$

which reduces to

$$\int_K \hat{f}(k^{-1}.w, 0)\, dk = (\hat{\mu}_{K.w}, f).$$

Therefore , we have $\varphi_w(z,t) = \hat{\mu}_{K.w}(z,0)$.

For $K = U(n)$, the distinct K-orbits are parametrised by $\tau \geq 0$. For $\tau = 0$, we have the trivial representation, and the $U(n)$ spherical function associated to that is $\eta_0(z,t) = 1$. For each $\tau > 0$ the sphere S_τ of radius τ in $\mathbb{C}^n$ is a K-orbit and the associated K-spherical function is $\eta_\tau = \hat{\mu}_\tau$ where μ_τ is the normalised surface measure on S_τ.

Proposition 3.2.5 *For each $\tau > 0$ we have a $U(n)$-spherical function*

$$\eta_\tau(z,t) = \frac{2^{n-1}(n-1)!}{(\tau|z|)^{n-1}} J_{n-1}(\tau|z|).$$

The proposition follows by explicitly calculating the Fourier transform of the surface measures μ_τ which are given by the Bessel function J_{n-1} of order $(n-1)$. To describe the $T(n)$-spherical functions associated to the representations σ_w, we make the following observation. If $K = K_1 \times K_2$ where $K_i \subset U(n_i)$, then the one point K-orbits are naturally parametrised by (w_1, w_2) where $w_i \in \mathbb{C}^{n_i}$ with

$$K.(w_1, w_2) = \{(\sigma_1 w_1, \sigma_2 w_2) : \sigma_i \in K_i\}.$$

One easily checks that

$$\varphi_{(w_1, w_2)}(z_1, z_2, t) = \varphi_{w_1}(z_1, t)\varphi_{w_2}(z_2, t)$$

and therefore, the $T(n)$ spherical functions are parametrised by $\rho \in (R^+)^n$ and are given by

$$\eta_\rho(z,t) = J_0(\rho_1|z_1|)J_0(\rho_2|z_2|).....J_0(\rho_n|z_n|).$$

Using properties of Bessel functions, we can show that these functions verify the defining equations for $T(n)$ spherical functions.

We now show that the functions E_k^λ and E_α^λ verify the defining conditions for elementary spherical functions. We first consider the functions E_k^λ.

Theorem 3.2.6 *For every $\lambda \neq 0$ and $k = 0, 1, \ldots$ we have*

$$\int_{U(n)} E_k^\lambda((z,t)(\sigma w, s)) \, d\sigma = E_k^\lambda(z,t)E_k^\lambda(w,s).$$

Proof: Recalling the definition of $E_k^\lambda(z,t)$, we only need to show that

$$\int_{U(n)} \varphi_k^\lambda(z + \sigma w)e^{\frac{i}{2}\lambda Im(z.\overline{\sigma w})} \, d\sigma$$

$$= \frac{k!(n-1)!}{(k+n-1)!}\varphi_k^\lambda(z)\varphi_k^\lambda(w).$$

Without loss of generality we can assume that $\lambda = 1$. By the bi-invariance of the Haar measure $d\sigma$ on $U(n)$, the value of the above integral depends only on $|z|$ and $|w|$. Therefore, it suffices to show that

$$\int_{|w|=r} \int_{U(n)} \varphi_k(z + \sigma w)e^{\frac{i}{2}Im(z.\overline{\sigma w})} \, d\sigma \, d\mu_r(w)$$

$$= \frac{k!(n-1)!}{(k+n-1)!}\varphi_k(z)\varphi_k(w).$$

But we already know from Theorem 2.4.4 that

$$\varphi_k \times \mu_r(z) = \frac{k!(n-1)!}{(k+n-1)!}\varphi_k(z)\varphi_k(r).$$

This proves the theorem.   ∎

In a similar fashion we can show that the functions E_α^λ satisfy the identity

$$\int_{T(n)} E_\alpha^\lambda((z,t)(e^{i\theta}w,s))\,d\theta = E_\alpha^\lambda(z)E_\alpha^\lambda(w).$$

Thus E_α^λ are indeed $T(n)$ spherical functions. The $U(n)$ spherical functions associated to the one dimensional representations are $\eta_\tau = \hat{\mu}_\tau$, and hence it is clear that

$$\eta_\tau * \mu_r(z) = \eta_\tau(r)\eta_\tau(z)$$

where $*$ stands for the convolution on $\mathbb{C}^n$. Using the above property it is easily verified that η_τ are indeed $U(n)$ spherical functions. Similarly, η_ρ can be shown to be $T(n)$ spherical functions. We omit the easy verifications.

The space $L^1(H^n)$ of integrable functions on H^n is a noncommutative Banach algebra under convolution. We are interested in the subalgebras $L^1(H^n/K)$ when $K = U(n)$ or $T(n)$. These are the subalgebras of $L^1(H^n)$ consisting of radial and polyradial functions respectively. As we show below, these subalgebras turn out to be commutative. This means that (H^n, K) is a Gelfand pair when $K = U(n)$ or $T(n)$. We identify the maximal ideal spaces of these subalgebras in terms of the associated spherical functions.

Lemma 3.2.7 *The algebras $L^1(H^n/U(n))$ and $L^1(H^n/T(n))$ are commutative.*

Proof: We consider the case of radial functions. It is enough to show that

$$(f * g)^\lambda(z) = (g * f)^\lambda(z)$$

for all λ and z. As $(f * g)^\lambda(z) = f^\lambda *_\lambda g^\lambda(z)$ it is enough to show that $f \times g = g \times f$ when f and g are radial functions on $\mathbb{C}^n$. But this follows from the fact that $W(f)$ and $W(g)$ commute as both are functions of the

Hermite operator. Similarly, one can show that the algebra of polyradial functions is commutative.   ∎

Let $A_r = L^1(H^n/U(n))$ be the Banach algebra of radial functions. If $M(A_r)$ stands for the maximal ideal space of the algebra A_r, then there is a one-to-one correspondence between the elements of $M(A_r)$ and the set of all nontrivial multiplicative linear functionals of A_r. Now any such homomorphism is of the form

$$\Lambda(f) = \int_{H^n} f(z,t)\varphi(z,t)\,dzdt$$

where φ is a bounded $U(n)$ spherical function. Therefore, we can identify $M(A_r)$ with the Gelfand spectrum $\Sigma(A_r)$ which by definition is the set of all bounded $U(n)$ spherical functions. The Gelfand spectrum $\Sigma(A_r)$ is the union of the Laguerre spectrum

$$\Sigma_L(A_r) = \{(\lambda,k) : \lambda \neq 0, k \in N\}$$

and the Bessel spectrum

$$\Sigma_B(A_r) = \{(0,\tau) : \tau \geq 0\}.$$

The Gelfand spectrum inherits the natural topology as a subset of the plane $\mathbb{R}^2$ which happens to be the Gelfand topology on the maximal ideal space. We denote points of $\Sigma(A_r)$ by the letter ζ and the associated spherical function will be denoted by φ_ζ. The Gelfand transform, denoted by $\mathcal{G}$, is defined to be the map which takes elements of A_r into continuous functions on $\Sigma(A_r)$ and is given by

$$\mathcal{G}f(\zeta) = \int_{H^n} f(z,t)\varphi_\zeta(z,t)\,dzdt.$$

Then $\mathcal{G}$ is a homomorphism of A_r into $C(\Sigma(A_r))$, the Banach algebra of continuous functions on the spectrum.

3.3   An algebra of radial measures

Consider the Heisenberg motion group $G = G^n$ acting on H^n. Let δ_g be the Dirac point mass at $g \in G$ and let m_K be the normalised Haar measure on $K = U(n)$. Let $*_G$ denote the convolution on G. An easy calculation shows that the measure

$$m_K *_G \delta_g *_G m_K, \quad g = (\sigma, z, t)$$

is independent of σ and depends only on $|z|$ and t. In fact, we have the relation

$$F *_G m_K *_G \delta_g *_G m_K(\tau, w, s)$$

$$= \int_K PF(w - kz, t + s - \frac{1}{2} Imkz.\bar{w}) \, dk$$

where we have written

$$PF(z, t) = \int_K F(k, z, t) \, dk.$$

We denote this measure by $m_{r,t}$ where $r = |z|$. If f is a function on the Heisenberg group, then treating it as a right K-invariant function on G we have

$$f *_G m_K *_G \delta_g *_G m_K(w, s) = f * \mu_{r,t}(w, s)$$

where $g = (e, z, t)$ and $r = |z|$. In the above $\mu_{r,t}$ stands for the measure on the sphere

$$S_{r,t} = \{(z, t) \in H^n : |z| = r\}.$$

When $t = 0$ we write $\mu_r = \mu_{r,0}$.

Fixing a unit vector $\omega \in \mathbb{C}^n$ and defining

$$G_\omega = \{g(r) = (e, r\omega, 0) \in G, r \in \mathbb{R}\},$$

we see that G_ω is a subgroup of G which is isomorphic to the group of reals $\mathbb{R}$. For $r > 0$ let m_r be the measure on G defined by

$$m_r = m_K *_G \delta_{g(r)} *_G m_K.$$

Note that for functions on H^n we have the relation

$$f *_G m_r(\sigma, z, t) = f * \mu_r(z, t).$$

In this section we consider the algebra generated by the measures m_r. This algebra is canonically isomorphic to the algebra generated by the measures μ_r on the Heisenberg group. We will now show that this algebra contains A_r. We start with a general result of Stempak [67] concerning the absolute continuity of convolution products of radial measures on H^n. Let $M(H^n)$ denote the algebra of finite measures on the Heisenberg group.

Theorem 3.3.1 *The convolution of two radial measures from $M(H^n)$ when $n \geq 2$ or three radial measures from $M(H^1)$ which do not have mass in the centre of H^n is absolutely continuous with respect to the Haar measure.*

To prove the theorem, we need to set up some notation and recall a result of Choquet which will be used in the proof. Let K_n stand for the set of all compactly supported rotation invariant probability measures on H^n. Let $ext(K_n)$ be the set of all extreme points of K_n. For each $r > 0$ and $t \in \mathbb{R}$, let $\mu_{r,t}$ be the normalised surface measure on the sphere

$$S_{r,t} = \{(z,t) \in H^n : |z| = r\}$$

and let δ_t be the Dirac mass concentrated at $(0,t)$. Then it is known that

$$ext(K_n) = E \cup \Delta_n$$

where $E = \{\mu_{r,t}\}$ and $\Delta_n = \{\delta_t\}$. For these facts see Stempak [67]. With these notations we have the following result due to Choquet: If μ is a compactly supported rotation invariant probabilty measure on H^n, then there exists a measure m_μ on $ext(K_n)$ such that

$$\mu(B) = \int_{ext(K_n)} \sigma(B) \, dm_\mu(\sigma).$$

This theorem is true in the general set up of locally compact topological groups. We refer to the paper of Ragozin [53]. Let us begin with the following lemma.

Lemma 3.3.2 *Let $\mu_j \in K_n$ for $j = 1, 2, \ldots, k$ and let M_j be the measures on $ext(K_n)$ associated to μ_j via Choquet's theorem. Then for any Borel subset B of H^n we have*

$$\mu_1 * \mu_2 * \ldots * \mu_k(B) = \int_{ext(K_n)} \cdots \int_{ext(K_n)}$$

$$\sigma_1 * \sigma_2 * \ldots * \sigma_k(B) \, dM_1(\sigma_1) dM_2(\sigma_2) \ldots dM_k(\sigma_k).$$

Proof: For $j = 1, 2, \ldots, k$ and a Borel set B, let $f(h_j)$ be the function defined by

$$f(h_j) = \int_{H^n} \cdots \int_{H^n} \chi_B(h_1 h_2 \ldots h_k)$$

$$\times d\mu_1(h_1) \ldots d\mu_{j-1}(h_{j-1}) d\mu_{j+1}(h_{j+1}) \ldots d\mu_k(h_k).$$

Then by the Choquet correspondence

$$\mu_1 * \mu_2 * \ldots * \mu_k(B)$$

$$= \int_{H^n} f(h_j) \, d\mu_j(h_j)$$

which is equal to

$$\int_{ext(K_n)} \sigma(f) \, dM_j(\sigma)$$

$$= \int_{ext(K_n)} \mu_1 * \ldots * \sigma * \ldots \mu_k(B) \, dM_j(\sigma).$$

Therefore, an easy induction proves the lemma.   ■

Proof of the theorem: It is enough to consider only probability measures. Since the measures have no mass at the centre, the above lemma gives

$$\mu_1 * \mu_2 * \ldots * \mu_k(B) = \int_E \ldots \int_E$$

$$\sigma_1 * \sigma_2 * \ldots * \sigma_k(B) \, dM_1(\sigma_1) dM_2(\sigma_2) \ldots dM_k(\sigma_k).$$

First we consider the case $n = 1$. Since $\mu_{r,t} = \mu_r * \delta_t$ and δ_t are in the centre of $M(H^1)$, it is enough to show that $\mu_p * \mu_q * \mu_r$ is absolutely continuous. Now μ_s is the measure on the circle

$$\{(s \cos \varphi, s \sin \varphi, 0) : 0 \le \varphi < 2\pi\}.$$

For $p, q, r > 0$ define the map $h : S^1 \times S^1 \times S^1 \to H^1$ by

$$h(\varphi, \psi, \theta) = (p \cos \varphi, p \sin \varphi, 0).(q \cos \psi, q \sin \psi, 0).(r \cos \theta, r \sin \theta, 0).$$

Then the Jacobian of this map is an analytic function of (φ, ψ, θ). It is easy to see that, at the point $(0, 0, \frac{\pi}{2})$, the Jacobian is nonzero.

Since the set of zeros of a nontrivial analytic function has measure zero, the above map has rank three except on a set D of measure zero. If $B \subset H^1$ has zero Haar measure then

$$\mu_p * \mu_q * \mu_r(B)$$

$$= \int_{S^1 \times S^1 \times S^1} \chi_B(h(\varphi, \psi, \theta)) \, dm(\varphi, \psi, \theta)$$

where dm is the Lebesgue measure on $S^1 \times S^1 \times S^1$. From the implicit function theorem, it follows that for every point at which h has rank three, there exists local coordinates in which h is identity. Since the measure m on $S^1 \times S^1 \times S^1$ is equivalent to the Lebesgue measure on any coordinate patch, from the above we deduce that

$$\mu_p * \mu_q * \mu_r(B) = 0.$$

This proves the absolute continuity of $\mu_p * \mu_q * \mu_r$.

Next consider the case $n \geq 2$. It is sufficient to show that $\mu_p * \mu_q$ is absolutely continuous. For $p, q > 0$ define a map

$$h : S^{2n-1} \times S^{2n-1} \to H^n$$

by $h(x,y) = (px,0).(qy,0)$ where $x, y \in S^{2n-1}$. We will show that h has rank $2n + 1$ everywhere except on a $\mu_1 \times \mu_1-$ null set D where μ_1 is the normalised surface measure on S^{2n-1}. As in the one dimensional case, it suffices to show that there exists a point at which h has rank $2n + 1$.

Consider the standard parametrisation near the point $(0, 0, .., 1)$ $\in S^{2n-1}$ given by

$$x = (x_1, x_2, \ldots, (1 - \sum_{j=1}^{2n-1} x_j^2)^{\frac{1}{2}}).$$

An easy computation shows that in these coordinates the Jacobian of h has rank $2n + 1$ at the point (x, y) where $x = (0, 0, \ldots, 1)$ and $y = (y_1, 0, \ldots, 0)$ with $|y_1|$ sufficiently small. Now for a Borel set B of measure zero, we have

$$\mu_p * \mu_q(B)$$

$$= \int_{S^{2n-1}} \int_{S^{2n-1}} \chi_B(h(x,y)) \, d\mu_1(x) d\mu_1(y)$$

which is nothing but

$$\mu_1 \times \mu_1(h^{-1}(B)) = \mu_1 \times \mu_1(h^{-1}(B) - D).$$

But at every point of $S^{2n-1} \times S^{2n-1} - D$, the implicit function theorem says that after a suitable change of coordinates, h is an orthogonal projection. By Fubini's theorem the inverse image under a projection of a Lebesgue null set is a null set and the measures $\mu_1 \times \mu_1$ and the Lebesgue measure in these coordinates are equivalent. So we get $\mu_p * \mu_q(B) = 0$ and this proves the absolute continuity. ∎

The measures $m_{r,t}$ on the Heisenberg motion group generate a closed Banach subalgebra of $M(G)$. Let us denote this algebra by B_r. Note that $m_{r,t}$ are $U(n)$- bi-invariant measures and so the algebra generated by them in $M(G)$ is canonically isomorphic to the algebra generated by $\mu_{r,t}$ in $M(H^n)$. We proceed to study the spectral theory of this Banach algebra.

Proposition 3.3.3 *Finite linear combinations of functions of the form*

$$\mu_{p,t} * \mu_{q,s} * \cdots * \mu_{r,u}$$

are dense in A_r and $A_r \subset B_r$.

Proof: By the result of Stempak [67] we already know that

$$\mu_r^{*3} = \mu_r * \mu_r * \mu_r$$

is an L^1 function, and hence $\varphi_r(z,t) = \mu_r^{*6}$ is a positive continuous function whose support is contained in

$$S_r^6 = \{h_1.h_2....h_6 : h_j \in S_r\}.$$

Here S_r stands for the sphere $\{(z,0) : |z| = r\}$. If δ_r stands for the Heisenberg dilations, it follows that

$$\delta_r \varphi_1(z,t) = \varphi_r(z,t).$$

Thus φ_r is a radial approximate identity and therefore, it follows from the theory of homogeneous nilpotent groups that the maximal operator

$$f \rightarrow \sup_{r>0} |f * \varphi_r|$$

satisfies a strong maximal inequality on $L^p(H^n), 1 < p < \infty$. For this fact see Folland-Stein [29]. Hence for every locally integrable function f on H^n, the convolution

$$f * \varphi_r(z,t) \rightarrow f(z,t)$$

for almost every (z,t). Integrating with respect to $\mu_{s,t}$, we see that

$$(\varphi_r, \mu_{s,t} * f) \rightarrow (\mu_{s,t}, f).$$

It follows that the space of functions spanned by convolutions of the spherical measures $\mu_{r,t}$ is dense in A_r. Indeed, every bounded linear functional on this space is given by integration against a bounded radial function f and if f satisfies

$$(\mu_{s,t} * \varphi_r, f) = 0$$

for all r, s and t, then as observed above $f = 0$ a.e. Since B_r is closed in the L^1 norm, it follows that $A_r \subset B_r$. ∎

We now proceed to prove that each complex homomorphism of B_r is given by a spherical function.

Proposition 3.3.4 *Let Λ be a nonzero continuous complex homomorphism of B_r. Then the radial function $\varphi(g) = \Lambda(m_g)$ equals $\bar{\varphi}_\zeta$ for some $\zeta \in \Sigma(A_r)$.*

Proof: As Λ is nontrivial, for some r, t we have $\Lambda(m_{r,t}) \neq 0$. Since $A_r \subset B_r$, Λ defines a homomorphism of A_r which is nontrivial as $\Lambda(m_{r,t}^{*3}) \neq 0$. Therefore, there is a bounded $U(n)$-spherical function φ_ζ such that

$$\Lambda(f) = \int_{H^n} f(z,t)\varphi_\zeta(z,t)\,dzdt$$

for all $f \in A_r$. We claim that $\varphi(g) = \bar{\varphi}_\zeta(g)$.

First we observe that

$$\Lambda(m_{r,t} * m_{r',t'})$$

$$= \int_{H^n} \varphi_\zeta(w,s)m_{r,t} * m_{r',t'}(w,s)\,dwds$$

as $m_{r,t} * m_{r',t'}$ is absolutely continuous. The above integral is equal to

$$\int_{H^n} \varphi_\zeta * m_{r',-t'}(w,s)\,dm_{r,t}$$

$$= \int_{H^n}\int_{H^n} \varphi_\zeta((z,t)(z',-t')^{-1})\,dm_{r',-t'}dm_{r,t}.$$

Since φ_ζ is $U(n)$-spherical, it follows that

$$\Lambda(m_{r,t} * m_{r',t'}) = \varphi_\zeta(z,t)\varphi_\zeta(z',-t')$$

where $|z| = r$ and $|z'| = r'$. On the other hand,

$$\Lambda(m_{r,t} * m_{r',t'}) = \varphi(z,t)\varphi(z',t')$$

since Λ is a homomorphism. Since we assume that $\varphi(z,t) \neq 0$, we see that $\varphi(z',t')$ is proportional to $\varphi_\zeta(z',-t')$ and hence they are equal since they agree at the origin. This concludes the proof of the proposition. ∎

Corollary 3.3.5 *Restriction of complex homomorphisms from B_r to its subalgebra A_r induces a canonical identification of the Gelfand spectrum of the two algebras.*

Proof: Since $m_{r,t}^{*3}$ is absolutely continuous, distinct nonzero homomorphisms of B_r restrict to distinct nonzero homomorphisms of A_r. Conversely, every continuous homomorphism of A_r is given by integration against a spherical function φ_ζ. Integrating φ_ζ against the radial measures $m_{r,t}$ defines a homomorphism of B_r. The multiplicative property follows from the functional equation satisfied by the spherical functions using a standard computation. The corollary is therefore proved. ∎

Now let π be a strongly continuous unitary representation of G on a Hilbert space $\mathcal{H}$. Then π determines canonically a norm continuous *-representation of the algebra $M(G//K)$ of K-bi-invariant measures on G. Let us denote by B_π the commutative C^* algebra which is the closure of $\pi(B_r)$ in the operator norm. Let Σ_π denote the spectrum of B_π which is by definition the set of all nonzero norm-continuous complex homomorphisms of B_π. Clearly, Σ_π is a subset of the Gelfand spectrum $\Sigma(B_r)$ of B_r. Consequently, every symmetric (self-adjoint) measure μ in $M(G//K)$ is mapped to a self-adjoint operator on $\mathcal{H}$ whose spectrum is the set $\{\mu(\varphi_\zeta) : \zeta \in \Sigma_\pi\}$. As a result we have the following formula which will be used in the study of maximal spherical means. If $f, g \in \mathcal{H}$ then

$$(\pi(\mu)f, g) = \int_{\Sigma_\pi} \mu(\varphi_\zeta) \, d\nu_{f,g}$$

where $d\nu_{f,g}$ is the spectral measure determined by the functions f and g.

When $\mu = m_r$ we have $\mu(\varphi_\zeta) = \varphi_\zeta(r)$ and we have the formula

$$\|\pi(m_r)f\|_2^2 = \int_{\Sigma_\pi} |\varphi_\zeta(r)|^2 \, d\nu_f$$

where we have written ν_f in place of $\nu_{f,f}$. More generally we have the following result.

Proposition 3.3.6 *For all smooth functions f with compact support and for all $k \geq 0$ we have*

$$\|\frac{d^k}{dr^k}(f * \mu_r)\|_2^2 = \int_{\Sigma_\pi} |\frac{d^k}{dr^k}\varphi_\zeta(r)|^2 \, d\nu_f(\zeta)$$

where $\nu_f(\zeta)$ is the spectral measure determined by f.

As the proof of this proposition is a bit technical we omit it. We refer to Nevo-Stein [48] and Nevo-Thangavelu [50] for a proof. In order to apply the above proposition to study the g_k functions, we need to get good estimates on the derivatives of the spherical functions. The next proposition gives the required estimates.

Proposition 3.3.7 *Let $n \geq 2$ and $0 \leq m \leq (n-1)$. Then*

$$\sup_{\zeta \in \Sigma} \int_0^\infty |\frac{d^m}{dr^m} \varphi_\zeta(r)|^2 r^{2m-1} \, dr \leq C_m(n)$$

where $C_m(n)$ depends only on m and n.

Proof: First consider the Laguerre spectrum. By making a change of variables, it is easy to see that we need to show that the following integrals

$$\left(\frac{k!}{(k+n-1)!}\right)^2 \int_0^\infty |\frac{d^m}{dr^m} \varphi_k(r)|^2 r^{2m-1} \, dr$$

are uniformly bounded for all $k \in N$. The Laguerre polynomials satisfy the formula

$$\frac{d}{dr} L_k^{n-1}(r) = -L_{k-1}^n(r)$$

(see 5.1.14 of Szego) and consequently we have

$$\frac{d}{dr} \varphi_k(r) = \left(-r L_{k-1}^n(\frac{r^2}{2}) - \frac{r}{2} L_k^{n-1}(\frac{r^2}{2})\right) e^{-\frac{1}{4}r^2}.$$

Iteration shows that the mth derivative of $\varphi_k(r)$ is a finite linear combination of terms of the form

$$r^{i+j} L_{k-j}^{n-1+j}(\frac{r^2}{2}) e^{-\frac{1}{4}r^2}$$

where $|i| \leq j$ and $j + |i| \leq m$. Let

$$\mathcal{L}_k^\alpha(r) = \left(\frac{\Gamma(k+1)}{\Gamma(k+\alpha+1)}\right)^{\frac{1}{2}} L_k^\alpha(r) r^{\frac{\alpha}{2}} e^{-\frac{r}{2}}$$

be the normalised Laguerre functions. The derivatives of $\varphi_k(r)$ can be expressed in terms of $\mathcal{L}_k^\alpha(r)$. Noting that

$$\frac{\Gamma(k+1)}{\Gamma(k+\alpha+1)} = O(k^\alpha)$$

we need to show that

$$k^{-n+j+1} \int_0^\infty |\mathcal{L}_{k-j}^{n-1+j}(r)|^2 r^{-(n-i-m)} \, dr \le C_m$$

where C_m is independent of k.

In order to estimate the foregoing integrals, we make use of the following lemma which is proved in [84] (see Lemma 1.5.4).

Lemma 3.3.8 *We let* $\alpha > -1$ *and* $\alpha + \beta > -1$. *Then* (i) $\|\mathcal{L}_k^{\alpha+\beta}(r)r^{-\frac{\beta}{2}}\|_2 = O(k^{-\frac{\beta}{2}})$ *when* $\beta < \frac{1}{2}$ *and* (ii) $\|\mathcal{L}_k^{\alpha+\beta}(r)r^{-\frac{\beta}{2}}\|_2 = O(k^{\frac{\beta-1}{2}})$ *when* $\beta > \frac{1}{2}$.

According to this lemma we need to treat two cases. First assume that $(n - i - m) > \frac{1}{2}$. Then we have

$$k^{-n+j+1} \int_0^\infty |\mathcal{L}_{k-j}^{n-1+j}(r)|^2 r^{-(n-i-m)} \, dr$$

$$\le C k^{-n+j+1} k^{n-i-m-1}.$$

Since $j + |i| \le m$ the desired estimate is established. When $(n - i - m) < \frac{1}{2}$ we have the estimate

$$k^{-n+j+1} \int_0^\infty |\mathcal{L}_{k-j}^{n-1+j}(r)|^2 r^{-(n-i-m)} \, dr$$

$$\le C k^{-n+j+1} k^{-(n-i-m)}$$

which is dominated by

$$C k^{-2n+i+j+m+1},$$

and as $j + |i| \le m$, we get the estimate $C k^{2(m-n)+1}$. The latter expression is bounded provided $m \le (n - \frac{1}{2})$ which holds by our assumption $m \le (n - 1)$.

Thus we have taken care of the Laguerre spectrum. On the Bessel spectrum we have to estimate derivatives of the Bessel function, η_τ. The Bessel function, $J_\alpha(s)$ satisfies the formula

$$\frac{d}{ds}(s^{-\alpha} J_\alpha(s)) = -s^{-\alpha} J_{\alpha+1}(s).$$

Using this and the asymptotic estimates for the Bessel functions, (which can be found e.g., in Watson [90]) we can estimate the integrals involving η_τ. The details are omitted. ∎

3.4 Analogues of Wiener-Tauberian theorem

Consider the following problem for the commutative Banach algebra $L^1(\mathbb{R}^n)$. Given $f \in L^1(\mathbb{R}^n)$, find conditions on f so that the closed subspace generated by f and all its translates $f(x + y)$, $y \in \mathbb{R}^n$ is the whole of $L^1(\mathbb{R}^n)$. The answer to this problem is the celebrated theorem of Wiener: the subspace is the whole of $L^1(\mathbb{R}^n)$ if and only if $\hat{f}(\xi)$ never vanishes. Here $\hat{f}(\xi)$ is the Euclidean Fourier transform of f. In the language of commutative Banach algebras, the above problem can be phrased as follows. Given a closed ideal J in $L^1(\mathbb{R}^n)$ under what conditions is $J = L^1(R^n)$? The answer can be rephrased in terms of the Gelfand transform: $J = L^1(\mathbb{R}^n)$ if and only if the following condition holds: given any $\xi \in \mathbb{R}^n$ there is a function $f \in J$ such that $\mathcal{G}f(\xi) \neq 0$. Note that the Gelfand transform for $L^1(\mathbb{R}^n)$ is precisely the Fourier transform.

Analogues of Wiener's theorem for L^p functions when p is different from 1 or 2 are difficult even in the case of $\mathbb{R}$; see Donoghue [18] Section 58. On the other hand, if we consider the action of the Euclidean motion group on $\mathbb{R}^n$, the corresponding problem has a nice answer. That is, instead of considering only translations, we consider translations and rotations. Given a function $f \in L^p(\mathbb{R}^n)$, we are interested in knowing when the smallest closed translation and rotation invariant subspace generated by f is the whole of $L^p(\mathbb{R}^n)$. Recently a complete answer to this problem was found by Rawat and Sitaram [56]. Again the conditions are in terms of the Gelfand transform.

In this section we are interested in analogues of Wiener's theorem for the Heisenberg group. Since we are in the noncommutative setup, the problem is more difficult. For the two-sided action of H^n on itself, a theorem of Wiener-Tauberian type has been known for some time; see Leptin [38]. However, if we consider Gelfand pairs (G^n, K) for which $L^1(H^n/K)$ is commutative, we can prove an analogue of Wiener's theorem. Below we will state a result of Hulanicki and Ricci [35] for the case $K = U(n)$. Instead of considering the action of H^n on itself, we can consider the action of G^n on H^n and look for an analogue of Wiener's theorem for translation and rotation invariant subspaces of $L^1(H^n)$. We prove one such result due to Rawat [55]. The conditions on the function are in terms of the Gelfand transform for the algebra $L^1(H^n/U(n))$.

We first consider the result of Hulanicki and Ricci [35]. From the general theory it is known that a symmetric Banach algebra A has the Wiener-Tauberian property if it satisfies the following three conditions:

(i) A is regular (ii) A contains an approximate identity and (iii) the set of all functions in A whose Gelfand transforms have compact support is dense in A. First we observe that the algebra $L^1(H^n/U(n))$ is symmetric. This means that for every complex homomorphism Λ of $L^1(H^n/U(n))$ one has $\Lambda(f^*) = \overline{\Lambda(f)}$ where $f^*(z,t) = \bar{f}(-z,-t)$. But this is clear since any such Λ is given by a bounded $U(n)$-spherical function, either E_k^λ or η_τ.

To verify the other conditions, we use Dixmier's functional calculus, [17]. If f is a self adjoint element of $L^1(H^n/U(n))$ then by this calculus there exists a C^k function F on $\mathbb{R}$ with $\int F(t)\,dt = 0$ such that

$$\int_R F(t)(e^{it\mathcal{G}(f)} - 1)\,dt = \mathcal{G}(g),$$

for some $g \in L^1(H^n/U(n))$. Using this one can show that the algebra $L^1(H^n/U(n))$ is regular. That is, the Gelfand topology on the maximal ideal space is Hausdorff. Also by choosing a compactly supported F with $F(1) = 1$ and $F(t) = 0$ for $|t| \leq \frac{1}{2}$, we can easily verify that $\mathcal{G}g = F(\mathcal{G}f)$ is compactly supported and g_s forms an approximate identity as $s \to 0$. Here $g_s(z,t) = s^{-Q}g(s^{-1}z, s^{-2}t)$. We now state the result of Hulanicki and Ricci.

Theorem 3.4.1 *Let J be a closed ideal in $L^1(H^n/U(n))$ and suppose that for every bounded $U(n)$-spherical function φ there is an $f \in J$ such that*

$$\int f(z,t)\varphi(z,t)\,dz\,dt \neq 0.$$

Then $J = L^1(H^n/U(n))$.

For details of the proof of this theorem, we refer to Faraut-Harzallah [21]. We now turn our attention to a Wiener-Tauberian theorem for the full algebra $L^1(H^n)$, but instead of considering only translations, we also consider rotations. For the rest of this section we use the notation $G = G^n$ and $K = U(n)$. Let f be a function on H^n. For $g \in G$ we define $f^g(z,t) = f(g.(z,t))$ to be the function translated by g. Given $f \in L^1(H^n)$, let $V(f)$ stand for the closed subspace spanned by $\{f^g : g \in G\}$. We want to know under what conditions on f the equality $V(f) = L^1(H^n)$ holds. The answer is given in the following theorem of Wiener-Tauberian type. Recall that $P_k(\lambda)$ is the projection of $L^2(\mathbb{R}^n)$ onto the kth eigenspace of the scaled Hermite operator $H(\lambda)$ which is spanned by $\{\Phi_\alpha^\lambda : |\alpha| = k\}$ and

$$\pi(f) = \int_{H^n} f(z,t)\pi(z,t)\,dz\,dt$$

for any representation π.

Theorem 3.4.2 *Let $f \in L^1(H^n)$. Then $V(f) = L^1(H^n)$ if and only if the following three conditions are satisfied: (i)$\pi_\lambda(f)^* P_k(\lambda) \neq 0$ for any $\lambda \neq 0$ and $k \in N$ (ii) for each $r > 0$ there is a $w \in \mathbb{C}^n$ with $|w| = r$ such that $\chi_w(f) \neq 0$ and (iii)$\int f(z,t)\,dz\,dt \neq 0$.*

Proof: We first prove the sufficiency of the conditions. As observed earlier, the given function f can be thought of as a right K-invariant function on G via $f((\sigma, z, t)) = f(z, t)$. Define another function f^* by setting

$$f^*((\sigma, z, t)) = \bar{f}((\sigma, z, t)^{-1}) = \bar{f}(-\sigma^{-1}z, -t).$$

Then f^* is a left K-invariant function. As can be easily checked, the function $f^* *_G f$ is a K-bi-invariant function on G. In the above $*_G$ stands for the convolution on the group G. Equivalently, $f^* *_G f$ can be viewed as a K-invariant function on H^n.

We claim that the closed ideal J generated by $f^* *_G f$ is the whole of $L^1(H^n/K)$. Assuming the claim for a moment, we will show how the theorem is proved. First we note that for any $h \in L^1(H^n/K)$, the function $h * (f^* *_G f)$ belongs to $V(f)$. So if the claim is true then it follows that

$$L^1(G//K) \subset V(f) \subset L^1(G/K) = L^1(H^n).$$

Thus $V(f)$ is a closed subspace of $L^1(G/K)$ which contains $L^1(G//K)$. Since it is also invariant under the left action of G, it has to be all of $L^1(G/K)$ which proves the theorem.

In order to prove the claim we apply Theorem 3.3.1 to the ideal J. Consider

$$\int_{H^n} f^* *_G f(z, t) E_k^\lambda(z, t)\,dz\,dt$$

$$= \frac{k!(n-1)!}{(k+n-1)!} \int_{H^n} f^* *_G f(z, t) e^{i\lambda t} \varphi_k^\lambda(z)\,dz\,dt.$$

As we have the relation

$$\varphi_k^\lambda(z) = (2\pi)^{\frac{n}{2}} \sum_{|\alpha|=k} (\pi_\lambda(z, 0)\Phi_\alpha^\lambda, \Phi_\alpha^\lambda)$$

we can write the above as

$$\int_{H^n} f^* *_G f(z, t) e^{i\lambda t} \varphi_k^\lambda(z)\,dz\,dt = (2\pi)^{\frac{n}{2}} \frac{k!(n-1)!}{(k+n-1)!}$$

$$\times \sum_{|\alpha|=k} \int_{H^n} f^* *_G f(z,t) (\pi_\lambda(z,t)\Phi_\alpha^\lambda, \Phi_\alpha^\lambda) \, dz dt.$$

Recalling the definition of the Fourier transform on H^n, the right hand side is given by the sum

$$(2\pi)^{\frac{n}{2}} \sum_{|\alpha|=k} (\pi_\lambda(f^* *_G f)\Phi_\alpha^\lambda, \Phi_\alpha^\lambda).$$

Now the convolution $f^* *_G f$ can be written in terms of the convolution on the Heisenberg group. It can be easily verified that

$$f^* *_G f(z,t) = \int_K f * f^*(\sigma z, t) \, d\sigma$$

where f^* on the right hand side stands for the function $\bar{f}(-z, -t)$. Therefore, we have

$$\pi_\lambda(f^* *_G f) = \int_K \pi_{\lambda,\sigma}(f)\pi_{\lambda,\sigma}(f)^* \, d\sigma$$

where $\pi_{\lambda,\sigma}(z,t) = \pi_\lambda(\sigma z, t)$. This gives us the relation

$$(\pi_\lambda(f^* *_G f)\Phi_\alpha^\lambda, \Phi_\alpha^\lambda) = \int_K \|\pi_{\lambda,\sigma}(f)^*\Phi_\alpha^\lambda\|_2^2 \, d\sigma.$$

Thus we see that

$$\int_{H^n} f^* *_G f(z,t) E_k^\lambda(z,t) \, dz dt = 0$$

if and only if

$$\pi_{\lambda,\sigma}(f)^*\Phi_\alpha^\lambda = 0$$

for all α with $|\alpha| = k$ and for almost all $\sigma \in K$.

But the representations $\pi_{\lambda,\sigma}$ and π_λ are equivalent via the intertwining operators $\mu_\lambda(\sigma)$ which is the metaplectic representation. These representations have the property that they map the eigenspace $span\{\Phi_\alpha^\lambda : |\alpha| = k\}$ onto itself. Therefore, $\pi_{\lambda,\sigma}(f)^*\Phi_\alpha^\lambda = 0$ if and only if $\pi_\lambda(f)^*\Phi_\alpha^\lambda = 0$. But this means that $\pi_\lambda(f)^*P_k(\lambda) = 0$. So we have proved that under the hypothesis (i) of the theorem

$$\int_{H^n} f^* *_G f(z,t) E_k^\lambda(z,t) \, dz dt \neq 0$$

for any λ and k.

Next consider the elementary spherical functions η_τ. From the definition, it follows that

$$\int_{H^n} f^* *_G f(z,t)\eta_\tau(z,t)\,dzdt$$

$$= \int_{H^n} \int_K f^* *_G f(z,t)e^{iRe(\sigma w).\bar{z}}\,d\sigma dzdt$$

where $w \in \mathbb{C}^n$ is such that $|w| = \tau$. Again using

$$f^* *_G f(z,t) = \int_K f * f^*(\sigma z,t)\,d\sigma$$

we get

$$\int_{H^n} f^* *_G f(z,t)\eta_\tau(z,t)\,dzdt$$

$$= C \int_K |\chi_{\sigma w}(f)|^2\,d\sigma$$

for a nonzero constant C. Therefore,

$$\int_{H^n} f^* *_G f(z,t)\eta_\tau(z,t)\,dzdt = 0$$

if and only if $\chi_{\sigma w}(f) = 0$ for all $\sigma \in K$. By the hypothesis (ii) this is not possible and hence

$$\int_{H^n} f^* *_G f(z,t)\eta_\tau(z,t)\,dzdt \neq 0$$

for any $\tau > 0$. Finally when $\tau = 0$, $\eta_0 = 1$ and

$$\int_{H^n} f^* *_G f(z,t)\,dzdt$$

$$= |\int_{H^n} f(z,t)\,dzdt|^2 \neq 0$$

by the hypothesis (iii). Hence all the conditions of Theorem 3.3.1 are verified and it follows that $J = L^1(H^n/K)$. This proves the claim.

Coming to the proof of necessity of the conditions, suppose any of them is violated by f. Then the same condition is violated by all functions in $V(f)$. If $V(f) = L^1(H^n/K)$ then any function in $L^1(H^n/K)$ should violate the same condition. But this is not the case. If the condition violated is

$$\int_{H^n} f(z,t)E_k^\lambda(z,t)\,dzdt \neq 0$$

then the function

$$h(z,t) = e^{-\frac{1}{2}t^2} \varphi_k^\lambda(z)$$

does not violate this. If the violated condition is

$$\int_{H^n} f(z,t)\eta_\tau(z,t)\, dzdt \neq 0$$

then the function

$$h(z,t) = e^{-\frac{1}{2}(|z|^2+t^2)}$$

does not violate the same . Hence $V(f)$ has to be a proper subspace of $L^1(H^n/K)$.   ∎

It would be interesting to see if there is an analogue of the above theorem for L^p functions. In the next chapter we will prove an L^p version for functions on the reduced Heisenberg group.

3.5   Spherical means on the Heisenberg group

Our aim in this section is to study the injectivity of the spherical mean value operator on the Heisenberg group using results proved in the previous section. A basic problem in integral geometry is to know when a continuous function f is uniquely determined by its averages over lower dimensional sets. For example, given a continuous function f on $\mathbb{R}^n$, we ask if it is uniquely determined by its averages over all spheres of a fixed radius r. In other words, if ν_r is the normalised surface measure on the sphere $|x| = r$, then we want to know if $f * \nu_r = 0$ implies $f = 0$. Here $f * \nu_r$ are called the spherical means of f for obvious reasons. An answer to this problem, in general, is in the negative. There are nontrivial bounded continuous functions f such that $f * \nu_r = 0$ for some r.

A counterexample is provided by the Bessel function

$$\varphi_\lambda(x) = c_n(\lambda|x|)^{-\frac{n}{2}+1} J_{\frac{n}{2}-1}(\lambda|x|)$$

where c_n is a suitable constant. Since $\varphi_\lambda(x)$ is the Fourier transform of the surface measure on the sphere $|x| = \lambda$, it satisfies the relation

$$\varphi_\lambda * \nu_r(x) = \varphi_\lambda(x)\varphi_\lambda(r).$$

Therefore, it is clear that all the averages of $\varphi_\lambda(x)$ vanish when $r\lambda$ is a zero of the Bessel function $J_{\frac{n}{2}-1}(t)$. On the other hand, if we consider averages over two families of spheres of radii r and s then f is uniquely determined provided the ratio r/s is not a quotient of zeros of $J_{\frac{n}{2}-1}(t)$. This is the so-called two radius theorem for $\mathbb{R}^n$.

We ask for the injectivity of the spherical means under some growth conditions on the function f. For example, when $f \in L^p(\mathbb{R}^n)$ with $1 \leq p \leq 2$ by taking the Fourier transform of the equation $f * \nu_r = 0$ and noting that $\hat{\nu}_r$ has only a countable number of zeros, we conclude that the spherical mean value operator is injective on $L^p(\mathbb{R}^n)$ for $1 \leq p \leq 2$. The counterexamples we have are all Bessel functions and they belong to $L^p(\mathbb{R}^n)$ for all $p > \frac{2n}{n-1}$. As it turns out, the spherical mean value operator is injective on $L^p(\mathbb{R}^n)$ for all $1 \leq p \leq \frac{2n}{n-1}$. We refer to [87] for a proof of this result.

In this section we are interested in such problems on the Heisenberg group. We consider μ_r as a measure on the sphere $S_r = \{(z,0) \in H^n : |z| = r\}$ and define the spherical means on H^n as the convolution $f * \mu_r$. Here the convolution is taken on the Heisenberg group, and we can ask if the operator taking f into $f * \mu_r$ is injective. As in the Euclidean case on bounded continuous functions, the above operator fails to be injective. A counterexample can be constructed in the following way. By considering functions of the form $f(z,t) = e^{it}g(z)$, the convolution equation $f * \mu_r = 0$ reduces to $g \times \mu_r = 0$. If we take $g(z) = \varphi_k(z)$ then we have

$$\varphi_k \times \mu_r(z) = \frac{k!(n-1)!}{(k+n-1)!}\varphi_k(z)\varphi_k(r)$$

and therefore, if we choose r so that $\frac{r^2}{2}$ is zero of the polynomial $L_k^{n-1}(t)$ then the function $f(z,t) = e^{it}\varphi_k(z)$ satisfies $f * \mu_r = 0$.

Functions of the form $f(z,t) = e^{it}\varphi_k(z)$ are continuous and bounded but they are not in $L^p(H^n)$ for any $p < \infty$. Therefore, we may expect the spherical mean value operator to be injective on $L^p(H^n)$ for $1 \leq p < \infty$. Using the Abel summability result of Strichartz proved in the previous chapter, we will show that this is indeed the case. Before proving this one radius theorem we consider a two radius theorem for bounded continuous functions. First we need two simple lemmas.

Lemma 3.5.1 *Let* $\varphi \in C^2(H^n) \cap L^1(H^n)$ *be such that* $\partial_t^2 \varphi \in L^1(H^n)$ *and* $\varphi(z,.)$ *is compactly supported on* $\mathbb{R}$ *for every* $z \in \mathbb{C}^n$. *Then the function*

$$\psi(z,t) = (2\pi)^{-1} \int_{-\infty}^{\infty} e^{-i\lambda t}\varphi(z,\lambda) \, d\lambda$$

satisfies $\psi^\lambda(z) = \varphi(z,\lambda)$ *and* $\psi \in L^1(H^n)$.

Proof: It is clear that $\psi^\lambda(z) = \varphi(z, \lambda)$ and so it is enough to show that $\tilde{\varphi} \in L^1(H^n)$ where $\tilde{\varphi}$ is the partial inverse Fourier transform of φ in the central variable. Integrating by parts we have

$$\int_{-\infty}^{\infty} e^{-i\lambda t}\partial_t^2\varphi(z, \lambda)\, d\lambda = (it)^2\tilde{\varphi}(z, t)$$

which gives us the estimate

$$|\tilde{\varphi}(z, t)| \le Ct^{-2}\int_{-\infty}^{\infty}|\partial_t^2\varphi(z, \lambda)|\, d\lambda.$$

Since $\tilde{\varphi}(z, t)$ is bounded as a function of t, the above estimate clearly shows that $\tilde{\varphi} \in L^1(H^n)$. Hence the lemma. ∎

Lemma 3.5.2 *Let μ be a rotation invariant compactly supported Radon measure on H^n and $\varphi \in L^1(H^n/U(n))$. Then $\mu * \varphi \in L^1(H^n/U(n))$ and $\mathcal{G}(\mu * \varphi) = \mathcal{G}(\mu)\mathcal{G}(\varphi)$.*

Proof: It is clear that $\mu * \varphi$ is radial and integrable. The identity

$$\mathcal{G}(\mu * \varphi) = \mathcal{G}(\mu)\mathcal{G}(\varphi)$$

means that

$$\int_{H^n} \mu * \varphi(a)\psi(a)\, da$$

$$= \left(\int_{H^n} \psi(a)\, d\mu(a)\right)\left(\int_{H^n} \varphi(a)\psi(a)\, da\right)$$

for every $U(n)$-spherical function ψ. As $\mu * \varphi = \varphi * \mu$ we have

$$\int_{H^n} \mu * \varphi(a)\psi(a)\, da$$

$$= \int_{H^n}\int_{H^n} \psi(a)\varphi(ab^{-1})\, d\mu(b)\, da$$

which after a change of variable gives the equation

$$\int_{H^n} \mu * \varphi(a)\psi(a)\, da$$

$$= \int_{H^n}\left(\int_{H^n} \varphi(a)\psi(ab)\, da\right) d\mu(b). \qquad (3.5.1)$$

Since φ and ψ are rotation invariant, we can write

$$\int_{H^n} \varphi(a)\psi(ab)\,da$$

$$= \int_{H^n} \int_{U(n)} \varphi(a)\psi(a\sigma.b)\,d\sigma\,da.$$

But as ψ is $U(n)$-spherical,

$$\int_{U(n)} \psi(a\sigma.b)\,d\sigma = \psi(a)\psi(b)$$

and using this in the equation (3.5.1) we get

$$\int_{H^n} \mu * \varphi(a)\psi(a)\,da$$

$$= \left(\int_{H^n} \psi(a)\,d\mu(a)\right)\left(\int_{H^n} \varphi(a)\psi(a)\,da\right)$$

which proves the lemma. ∎

We are now ready for the two radius theorem which will follow as a corollary of the next result.

Theorem 3.5.3 *Let $\mathcal{R}$ be a family of rotation invariant compactly supported Radon measures on H^n such that, for any $U(n)$-spherical function ψ, there is a $\mu \in \mathcal{R}$ such that $\int_{H^n} \psi(a)\,d\mu(a) \neq 0$. If f is a continuous function such that $f * \mu = 0$ for all $\mu \in \mathcal{R}$, then $f = 0$.*

Proof: The condition $f * \mu = 0$ clearly implies that $f * \mu * \varphi = 0$ for any $\mu \in \mathcal{R}$ and $\varphi \in L^1(H^n/U(n))$. So the closed ideal J in $A_r = L^1(H^n/U(n))$ generated by

$$\{\mu * \varphi : \mu \in \mathcal{R}, \varphi \in A_r\}$$

satisfies $f * J = 0$. Therefore, in order to show that $f = 0$ it is enough to prove $J = A_r$. We prove this by appealing to the Wiener-Tauberian theorem proved in the previous section. We need to verify that given any $\Lambda \in M(A_r)$ there is an element $\eta \in J$ such that $\Lambda(\eta) \neq 0$. In view of the hypothesis on $\mathcal{R}$ and Lemma 3.5.2, it is enough to show that given any $\Lambda \in M(A_r)$ there is $\varphi \in A_r$ such that $\Lambda(\varphi) \neq 0$.

Let ψ be the spherical function associated to Λ. First assume that $\psi = E_k^\lambda$. Without loss of generality we can take $\lambda > 0$ and consider the function

$$\eta(z,t) = \int_{-\infty}^{\infty} e^{-ist}\varphi(s)\varphi_k^s(z)\,ds$$

where $\varphi \in C_0^\infty(\frac{\lambda}{2}, 2\lambda)$ with $\varphi(\lambda) = 1$. By Lemma 3.5.1 this function η belongs to A_r and

$$\eta^s(z) = \varphi(s)\varphi_k^s(z).$$

Hence $\Lambda(\eta)$ equals $\frac{k!(n-1)!}{(k+n-1)!}$ times the integral

$$\int_{\mathbb{C}^n} \eta^\lambda(z)\varphi_k^\lambda(z)\, dz = \varphi(\lambda)\|\varphi_k^\lambda\|_2^2$$

which is clearly nonzero. If the spherical function associated to Λ is $\eta_\tau(z,t)$, then we take

$$\eta(z,t) = e^{-\frac{1}{2}(|z|^2+t^2)}.$$

In this case, a simple calculation shows that

$$\Lambda(\eta) = c_n \tau^{1-n} \int_0^\infty e^{-\frac{1}{2}r^2} J_{n-1}(\tau r) r^n \, dr$$

which is again nonzero. Thus we can apply the Wiener-Tauberian theorem to the ideal J to conclude that $J = A_r$. This proves the theorem. ∎

Corollary 3.5.4 *Assume that f is a bounded continuous function which satisfies $f * \mu_r = f * \mu_s = 0$. Then $f = 0$ provided (i) $\frac{r^2}{s^2}$ is not a quotient of zeros of any $L_k^{n-1}(t)$ and (ii) $\frac{r}{s}$ is not a quotient of zeros of $J_{n-1}(t)$.*

Proof: We apply the theorem with $\mathcal{R} = \{\mu_r, \mu_s\}$. Suppose $\Lambda \in M(A_r)$ is given by the spherical function E_k^λ. Then a calculation shows that

$$\Lambda(\mu_t) = \frac{k!(n-1)!}{(k+n-1)!}\varphi_k^\lambda(t).$$

Since $\frac{r^2}{s^2}$ is not a quotient of zeros of $L_k^{n-1}(t)$ it follows that either $\Lambda(\mu_r)$ or $\Lambda(\mu_s)$ is non zero. If Λ is given by η_τ then

$$\Lambda(\mu_t) = c_n(t\tau)^{-n+1} J_{n-1}(t\tau)$$

and again one of $\Lambda(\mu_r)$ or $\Lambda(\mu_s)$ is nonzero. The corollary follows from the theorem. ∎

An analogue of the above theorem can be proved for families of compactly supported polyradial measures. Similarly, we can also prove two radius theorem for averages over tori

$$\{(z,0) : |z_j| = r_j, j = 1, 2, \ldots, n\}.$$

The formulation and proofs are left to the interested reader; see [1]. We now turn our attention to a one-radius theorem. First we prove a one-radius theorem for functions in $L^p(H^n)$ with $1 \leq p \leq 2$. The proof given for this case does not require the summability theorem of Strichartz and the proof can be adapted to prove a two-radius theorem for functions on the reduced Heisenberg group that are not in $L^p(H^n)$ for any p. Finally we will prove the one-radius theorem for $L^p(H^n)$ with $2 \leq p < \infty$.

Before stating the next proposition, let us set up some notation. Recall that K_n stands for the set of all compactly supported rotation invariant probability measures on H^n and $ext(K_n)$ for the set of all extreme points of K_n, so that $ext(K_n) = E \cup \Delta_n$ where $E = \{\mu_{r,t}\}$ and $\Delta_n = \{\delta_t\}$. If μ is a compactly supported rotation invariant probability measure on H^n let m_μ be the measure on $ext(K_n)$ associated to μ via the Choquet theorem. Finally given a finite measure μ let $\mu(e_k^\lambda)$ denote the average

$$\mu(e_k^\lambda) = \int_{H^n} e_k^\lambda(z,t) \, d\mu(z,t).$$

Proposition 3.5.5 *Assume that μ is a compactly supported rotation invariant probability measure with no mass at the centre of H^n. Then*

$$e_k^\lambda * \mu(z,t) = \frac{k!(n-1)!}{(k+n-1)!} \mu(e_k^\lambda) e_k^\lambda(z,t).$$

Proof: Since we are assuming that μ has no mass at the centre

$$Z = \{(0,t) : t \in \mathbb{R}\}$$

of the Heisenberg group, $\mu(Z) = 0$, which means that the measure m_μ is concentrated on E alone. Thus we have

$$e_k^\lambda * \mu(z,t) = \int_E e_k^\lambda * \sigma(z,t) \, dm_\mu(\sigma).$$

When $\sigma = \mu_{r,t}$, it is easy to calculate $e_k^\lambda * \sigma$. In fact,

$$e_k^\lambda * \mu_{r,t}(z,s)$$

$$= e^{i\lambda(s-t)} \int_{|w|=r} e^{\frac{i}{2}\lambda Im z.\bar{w}} \varphi_k^\lambda(z-w) \, d\mu_r.$$

We have already calculated this integral and we have

$$e_k^\lambda * \mu_{r,t}(z,s)$$

$$= \frac{k!(n-1)!}{(k+n-1)!} e_k^\lambda(r,t) e_k^\lambda(z,s)$$

where $e_k^\lambda(r,t)$ stands for $e_k^\lambda(w,t)$ with $|w| = r$. Using this in the above integral for μ we get the proposition. ∎

The following observation will be used in the proof of the one-radius theorem. As μ is compactly supported, it follows from the definition that $\mu(e_k^\lambda)$ extends to a holomorphic function of λ in the cut plane $Re\lambda \neq 0$. Therefore, $\mu(e_k^\lambda)$ can have at most countably many zeros for each k. Also $\mu(e_k^\lambda)$ is a continuous function of λ and

$$\lim_{\lambda \to 0} \mu(e_k^\lambda) = \frac{k!(n-1)!}{(k+n-1)!}$$

and hence there is a neighbourhood of 0 in which $\mu(e_k^\lambda)$ has no zeros. We are now ready for one-radius theorems.

Theorem 3.5.6 *Let μ be a compactly supported rotation invariant Radon measure with no mass at the centre of the Heisenberg group. If $f \in L^p \cap C(H^n)$, $1 \leq p \leq 2$ satisfies $f * \mu = 0$, then $f = 0$.*

Proof: Taking an approximate identity $g_n \in C_0^\infty(H^n)$, we have $g_n * f \to f$ in L^p. Therefore, it is enough to show that $g_n * f = 0$ for each n. As $g_n * f \in L^2(H^n)$ and $g_n * f * \mu = 0$, we can assume that $f \in L^2(H^n)$ to start with. Taking partial Fourier transform in the t variable, we get the equation $f^\lambda *_\lambda \mu^\lambda(z) = 0$ for almost all λ. We will show that this implies $f^\lambda = 0$ for almost every λ which will then prove the theorem.

In order to prove this, we expand f^λ in terms of special Hermite functions:

$$f^\lambda(z) = (2\pi)^{-n} \lambda^n \sum_{k=0}^\infty f^\lambda *_\lambda \varphi_k^\lambda(z).$$

The above series converges in the L^2 norm. To show that $f^\lambda = 0$, it is enough to show that $f^\lambda *_\lambda \varphi_k^\lambda = 0$ for each k. Taking the λ-twisted convolution of the equation $f^\lambda *_\lambda \mu^\lambda(z) = 0$ with φ_k^λ and using the result of the previous proposition, we get

$$\sum_{k=0}^\infty \frac{k!(n-1)!}{(k+n-1)!} \mu(e_k^\lambda) f^\lambda *_\lambda \varphi_k^\lambda(z) = 0.$$

Since $\mu(e_k^\lambda)$ vanishes only for a countable number of values of λ for each k, we see that $f^\lambda *_\lambda \varphi_k^\lambda = 0$ for almost every λ for each k. This proves our claim and the theorem follows. ∎

Using the approximation results proved in Section 1.4, we will now relax the condition on f in the z variable and prove a one-radius theorem. We say that f is of tempered growth if it defines a tempered distribution.

Theorem 3.5.7 *Let μ be as in the previous theorem. Assume that f is a continuous function on H^n, $f(z,.) \in L^p(\mathbb{R})$, $1 \le p \le 2$, and f^λ is of tempered growth on $\mathbb{C}^n$ for almost all λ. Then $f * \mu = 0$ implies $f = 0$.*

Proof: The integrability condition in the t variable allows us to take the partial Fourier transform of the equation $f * \mu = 0$ which gives $f^\lambda *_\lambda \mu^\lambda = 0$ for almost every λ. As before, we only need to show that $f^\lambda = 0$ for almost every λ. Since f^λ is of tempered growth, it is enough to show that $\int f^\lambda(z) g(z)\, dz = 0$ for every g in the Schwartz class. By Proposition 1.4.5 we know that the series $g = \sum_m R_m g$ converges in the Schwartz topology. Therefore, we can assume that g has some homogeneity.

As $f^\lambda(z)$ is a continuous function of z, we can form the Fourier series

$$f^\lambda(e^{i\theta}z) = \sum_m R_m f^\lambda(z) e^{im.\theta}$$

and therefore, in order to prove $f^\lambda = 0$, it is enough to show that $R_m f^\lambda(z) = 0$ for all m and z. Thus we need only to prove

$$\int R_m f^\lambda(z) g(z)\, dz = 0$$

for all homogeneous Schwartz class functions. By Proposition 1.4.6 we know that the special Hermite expansion

$$g(z) = (2\pi)^{-n} \lambda^n \sum_{k=0}^{\infty} g *_\lambda \varphi_k^\lambda(z)$$

converges in the Schwartz topology and therefore, we are led to show that

$$\int_{\mathbb{C}^n} R_m f^\lambda(z) g *_\lambda \varphi_k^\lambda(z)\, dz = 0$$

for every k. The above is equivalent to

$$\int_{\mathbb{C}^n} \varphi_k^\lambda *_\lambda R_m f^\lambda(z) g(z)\, dz = 0.$$

We claim that $\varphi_k^\lambda *_\lambda R_m f^\lambda(z) = 0$ for every k.

To prove the claim, we first observe that $\varphi_k^\lambda *_\lambda R_m f^\lambda(z)$ is a finite sum of special Hermite functions as $R_m f^\lambda$ is homogeneous. Consequently, $\varphi_k^\lambda *_\lambda R_m f^\lambda$ belongs to $L^2(H^n)$. Further, we also have

$$\varphi_k^\lambda *_\lambda R_m f^\lambda *_\lambda \mu^\lambda(z) = 0$$

which follows from the relation

$$R_m f^\lambda *_\lambda \mu^\lambda(z) = R_m(f^\lambda *_\lambda \mu^\lambda).$$

The later relation holds since μ is rotation invariant. Thus $\varphi_k^\lambda *_\lambda R_m f^\lambda$ is an L^2 function satisfying

$$\varphi_k^\lambda *_\lambda R_m f^\lambda *_\lambda \mu^\lambda(z) = 0.$$

Therefore, as in the previous theorem, we can conclude that $\varphi_k^\lambda *_\lambda R_m f^\lambda = 0$ for all k which proves our claim.

Thus we have shown that

$$\int_{\mathbb{C}^n} R_m f^\lambda(z) g *_\lambda \varphi_k^\lambda(z)\, dz = 0$$

for all k, which means that $R_m f^\lambda$ is orthogonal to all Schwartz class functions and hence $R_m f^\lambda = 0$. Since this is true for every m, we conclude that $f^\lambda = 0$ for almost every λ which proves the theorem. ∎

We conclude this section by proving the following one-radius theorem for functions in L^p with $1 \leq p < \infty$. It is in the proof of this theorem that we use the summability result proved in Chapter 2.

Theorem 3.5.8 *Let μ be a compactly supported rotation invariant probability measure with no mass at the centre of H^n. If f belongs to $L^p(H^n) \cap C(H^n)$, $1 \leq p < \infty$, and satisfies $f * \mu = 0$, then $f = 0$.*

Proof: It is enough to consider $p > 1$. Let T_a^b stand for the multiplier transform in the t variable corresponding to the multiplier $\chi_{(a,b)}(\lambda)$. Then by Corollary 2.2.6, we know that the functions

$$f_N = \sum_{j=0}^{N^2} (1 - \frac{1}{N})^j T_{-N}^N R_j f \qquad (3.5.2)$$

converge to f in the L^p norm as N tends to ∞. Since convolution with μ is a bounded operator on L^p and $f * \mu = 0$, we have $\lim_{N \to \infty} f_N * \mu = 0$.

In view of (3.5.2), we have

$$\lim_{N\to\infty} \sum_{j=0}^{N^2} (1 - \frac{1}{N})^j (T_{-N}^N R_j f) * \mu = 0.$$

By the result of Proposition 3.4.5, this is the same as saying

$$\lim_{N\to\infty} \sum_{j=0}^{N^2} (1 - \frac{1}{N})^j \int_{-N}^{N} \mu(e_j^{-\lambda})(f * e_j^{\lambda})|\lambda|^n \, d\lambda = 0.$$

Applying the operator $T_a^b R_k$ to the above sequence of functions, we conclude that

$$\int_a^b \mu(e_k^{-\lambda})(f * e_k^{\lambda})|\lambda|^n \, d\lambda = 0.$$

If a and b are such that for $a \leq \lambda \leq b$, $\mu(e_k^{-\lambda})$ never vanishes, then it is easy to see that $\chi_{(a,b)}(\lambda)(\mu(e_k^{-\lambda}))^{-1}$ is an $L^p(\mathbb{R})$ multiplier, and hence for such a and b, we get

$$\int_a^b (f * e_k^{\lambda}) |\lambda|^n d\lambda = 0.$$

But this means that the Fourier transform of $R_k f$ in the t variable is supported on the zero set of $\mu(e_k^{-\lambda})$ which is discrete. Therefore, as a function of t, $R_k f$ is an infinite sum of functions of the form $t^{m_j} e^{i\lambda_j t}$ which contradicts the hypothesis that $f \in L^p$ unless $R_k f = 0$. This shows that $R_k f = 0$ for all k and hence $f = 0$. This completes the proof of the theorem.   ∎

3.6   A maximal theorem for spherical means

The Lebesgue differentiation theorem states that when $f \in L^1(\mathbb{R}^n)$, its averages over the balls $B_r(x) = \{y : |x - y| \leq r\}$ converge to $f(x)$ as r tends to zero for almost all x. There is an analogue of this result for averages over spheres. Let $f * \nu_r(x)$ be the spherical means of a function on $\mathbb{R}^n$ and let

$$M_\nu f(x) = \sup_{r>0} |f * \nu_r(x)|$$

be the associated maximal function. Then a celebrated theorem of Stein [64] states that, in dimensions greater than two, M_ν is bounded on $L^p(\mathbb{R}^n)$ for all $p > \frac{n}{n-1}$. As a corollary, we get the almost everywhere convergence of the spherical means to the function. The same result holds in two dimensions as well, as was later proved by Bourgain [9]. In this section we are interested in proving a maximal theorem for the spherical means on the Heisenberg group.

Consider the spherical means $f * \mu_r$ on the Heisenberg group studied in the previous section. Define the maximal function associated to these spherical means by

$$M_\mu f(z, t) = \sup_{r>0} |f * \mu_r(z, t)|.$$

In order to study the boundedness properties of this maximal function, we use square function arguments, analytic interpolation and the spectral theory developed in Section 3.4. We also need a result of Birkhoff [8] from ergodic theory. We first embed the spherical mean value operator in an analytic family of operators. The embedding is implemented by using the Riemann-Liouville fractional integrals.

Given a smooth function f on H^n, let

$$F_h(r) = f * \mu_r(h)$$

where $h = (z, t) \in H^n$. For $\alpha = a + ib$, the Riemann-Liouville fractional integrals of $F_h(r)$ are defined by

$$I^\alpha F_h(r) = \frac{1}{\Gamma(\alpha)} \int_0^r (r - s)^{\alpha-1} F_h(s)\, ds.$$

The normalised fractional integrals are then defined by

$$M^\alpha F_h(r) = r^{-\alpha} I^\alpha F_h(r).$$

These fractional integrals can be analytically continued to entire functions. For example, when $-1 < \alpha \le 0$, we write

$$I^\alpha F_h(r) = \frac{1}{\Gamma(\alpha)} \int_0^{\frac{r}{2}} (r - s)^{\alpha-1} F_h(s)\, ds$$

$$-\frac{1}{\alpha\Gamma(\alpha)} \int_{\frac{r}{2}}^r \frac{d}{ds} (r - s)^\alpha F_h(s)\, ds.$$

Integrating by parts we get the expression

$$I^\alpha F_h(r) = \frac{1}{\Gamma(\alpha)} \int_0^{\frac{r}{2}} (r - s)^{\alpha-1} F_h(s)\, ds$$

$$+\frac{1}{\Gamma(1+\alpha)}\left(\left(\frac{r}{2}\right)^\alpha F_h(\frac{r}{2}) + \int_{\frac{r}{2}}^r (r-s)^\alpha \frac{d}{ds} F_h(s)\, ds\right).$$

We take this to be the definition of $I^{-\alpha}F_h(r)$ when $-1 < \alpha \le 0$. Similarly, integrating by parts more and more, we can extend the definition to all negative values of α.

The family of operators I^α satisfy the functional equations

$$I^\alpha I^\beta F = I^{\alpha+\beta} F, \qquad I^0 F = F.$$

Clearly, by definition

$$I^1 F_h(r) = \int_0^r F_h(s)\, ds.$$

It follows that

$$I^{-1} F_h(r) = \frac{d}{dr} F_h(r)$$

and more generally

$$M^{-k} F_h(r) = r^k \frac{d^k}{dr^k}(f * \mu_r)(h).$$

We define the maximal operator

$$S_\alpha^* f(h) = \sup_{r>0} |M^\alpha F_h(r)|.$$

For these maximal operators we first prove the following estimate.

Proposition 3.6.1 *Let $1 < p < \infty$ and let $f \in L^p(H^n)$. Then we have*

$$\|S_{1+ib}^* f\|_p \le Ce^{\pi|b|}\|f\|_p.$$

Proof: Let

$$\nu_r = \frac{1}{r}\int_0^r \mu_s\, ds$$

be the uniform averages of the measures μ_s and consider the associated maximal function

$$M_\nu f(h) = \sup_{r>0} |f * \nu_r(h)|.$$

From the definition it is clear that $S_1^* f(h) \le CM_\nu f(h)$ and therefore, the estimate $\|S_1^* f\|_p \le C\|f\|_p$ would follow from the corresponding estimate for the maximal function $M_\nu f$. The operators S_{1+ib}^* satisfy the desired inequality as a standard consequence of Γ-function estimates. We need to show that M_ν is bounded on $L^p(H^n)$ for $1 < p < \infty$.

Consider the Heisenberg motion group G acting on itself by right translations. This action of G gives rise to a natural isometric representation of G on $L^p(G), 1 \leq p \leq \infty$ given by $\pi(g)f(h) = f(hg)$. This action is ergodic in the sense that there are no G-invariant functions in $L^2(G)$. In terms of this representation we can write the spherical means as

$$f * \mu_r(h) = \int_G \pi(g)f(h)\,dm_{g(r)}$$

where we have written

$$m_{g(r)} = m_K *_G \delta_{g(r)} *_G m_K$$

with $g(r) \in G_\omega$. In the above the function f is identified with a right $U(n)$-invariant function on G. We recall that G_ω is a subgroup of G which is isomorphic to the group $\mathbb{R}$ of real numbers. Thus we have

$$f * \mu_r = \pi(m_{g(r)})f = \pi(m_K)\pi(\delta_{g(r)})\pi(m_K)f.$$

From the above formula for the spherical means, it follows that

$$f * \nu_r(h)$$

$$= \pi(m_K)\left(\frac{1}{r}\int_0^r \pi(\delta_{g(s)})\,ds\right)\pi(m_K)f(h).$$

Here $\frac{1}{r}\int_0^r \pi(\delta_{g(s)})\,ds$ are the Birkhoff averages over the group $G_\omega = \mathbb{R}$. Therefore, by the strong maximal inequality for $\mathbb{R}$-actions, we know by a theorem of Birkhoff [8] that

$$\sup_{r>0} \left|\frac{1}{r}\int_0^r \pi(\delta_{g(s)})f(h)\,ds\right|$$

is bounded on $L^p(H^n), 1 < p < \infty$. As $\pi(m_K)$ is a positive projection bounded on $L^p(H^n)$, it follows that the same is true of M_ν. This completes the proof of the proposition. ∎

Next we establish an L^2 estimate for S_α^* when $Re(\alpha)$ is negative. To do this, we use square function arguments. For $k = 1, 2, \ldots$, we define the g_k functions by

$$g_k(f,h)^2 = \int_0^\infty |F_h^{(k)}(r)|^2 r^{2k-1}\,dr.$$

We first show that these functions are bounded on $L^2(H^n)$.

Proposition 3.6.2 *The estimates* $\|g_k(f)\|_2 \leq C_k\|f\|_2$ *hold for all* $1 \leq k \leq n - 1$.

Proof: From Proposition 3.3.6 it follows that

$$\|g_k(f)\|_2^2$$

$$= \int_\Sigma \int_0^\infty |\varphi_\zeta^{(k-1)}(r)|^2 r^{2k-1} \, dr d\nu_f(\zeta)$$

where ν_f is the spectral measure determined by f. According to Proposition 3.3.7 we have

$$\sup_{\zeta \in \Sigma} \int_0^\infty |\varphi_\zeta^{(k-1)}(r)|^2 r^{2k-1} \, dr \leq C.$$

The proposition immediately follows from this estimate.   ∎

The estimates on the g-functions are used in the proof of the following proposition.

Proposition 3.6.3 *Let* $n \geq 2$ *and* $1 \leq k \leq (n-1)$. *Then the maximal inequality*

$$\|S^*_{-k+1}f\|_2 \leq C\|f\|_2$$

holds for all $f \in L^2(H^n)$.

Proof: Integration by parts gives the relation

$$r^k F_h^{(k-1)}(r) = \int_0^r s^k F_h^{(k)}(s) \, ds$$

$$+ \int_0^r k s^{k-1} F_h^{(k-1)}(s) \, ds.$$

Dividing both sides by r and applying the Cauchy-Schwarz inequality to the integrals on the right, we get

$$|S^*_{-k+1}f(h)| \leq g_k(f, h) + C_k g_{k-1}(f, h).$$

The g-function estimates of the proposition complete the proof.   ∎

In the next proposition we obtain an improvement of this result by allowing fractional derivatives.

Proposition 3.6.4 *Let $n \geq 2$ and $\alpha = a + ib$. Assume that*

$$\|S^*_{-m+1}f\|_2 \leq C\|f\|_2.$$

Then the maximal inequality

$$\|S^*_\alpha f\|_2 \leq Ce^{\pi|b|}\|f\|_2$$

holds whenever $a > -m + \frac{1}{2}$.

Proof: Let $0 < \delta < \frac{1}{2}$ and consider $I^{-\delta-m+1}F_h(r)$. By the semigroup property of I^α, we have

$$I^{-\delta-m+1}F_h(r) = I^{-\delta}I^{-m+1}F_h(r).$$

Recalling the definition of I^α when $-1 < \alpha \leq 0$, we have the formula

$$I^{-\delta-m+1}F_h(r) = \frac{1}{\Gamma(-\delta)}\int_0^{\frac{r}{2}}(r-s)^{-\delta-1}I^{-m+1}F_h(s)\,ds$$

$$+\frac{1}{\Gamma(1-\delta)}\left((\frac{r}{2})^{-\delta}I^{-m+1}F_h(\frac{r}{2}) + \int_{\frac{r}{2}}^r(r-s)^{-\delta}\partial_s I^{-m+1}F_h(s)\,ds\right).$$

Recalling that $\frac{d}{ds} = I^{-1}$ and multiplying the above by $r^{\delta+m-1}$, the above can be written in the form

$$I^{-\delta-m+1}F_h(r) = \frac{1}{\Gamma(-\delta)}\int_0^{\frac{1}{2}}(1-s)^{-\delta-1}M^{-m+1}F_h(sr)\,ds$$

$$+\frac{1}{\Gamma(1-\delta)}\left(2^\delta M^{-m+1}F_h(\frac{r}{2}) + \int_{\frac{r}{2}}^r(r-s)^{-\delta}r^{\delta-\frac{1}{2}}r^{m-\frac{1}{2}}I^{-m}F_h(s)\,ds\right).$$

Now the first two terms are dominated by a constant times $S^*_{-m+1}f$. By the Cauchy-Schwarz inequality, the third term is bounded by (as $r \leq 2s$), a constant times

$$\left(\int_{\frac{r}{2}}^r(r-s)^{-2\delta}r^{2\delta-1}\,ds\right)^{\frac{1}{2}}\left(\int_{\frac{r}{2}}^r s^{2m-1}|F_h^{(m)}(s)|^2\,ds\right)^{\frac{1}{2}}.$$

Since $0 < \delta < \frac{1}{2}$ the above is dominated by $g_m(f,h)^2$. Therefore, if $S^*_{-m+1}f$ is bounded on L^2, then the same is true of $S^*_a f$ for $a > -m+\frac{1}{2}$. By using estimates for the gamma function, we conclude that the same is true of $S^*_\alpha f$ when $Re(\alpha) > -m + \frac{1}{2}$. This proves the proposition. ∎

We can now state and prove the following maximal theorem for spherical means on the Heisenberg group.

Theorem 3.6.5 *Let $\frac{2n-1}{2n-2} < p < \infty$ and $n \geq 2$. Then the maximal function $M_\mu f$ associated to the spherical means is bounded on $L^p(H^n)$. Moreover, $f * \mu_r(h)$ converges to $f(h)$ for almost every h as r tends to zero.*

Proof: The proof follows by analytic interpolation. We apply the analytic interpolation theorem of Stein to the family S_α^*. The end point estimates are provided by the Propositions 3.6.1 and 3.6.4. Some technical difficulties arise owing to the fact that S_α^* are not linear. For details we refer to Nevo-Stein [49]. If f is a continuous function then $f * \mu_r$ clearly converges to f pointwise. With maximal inequality in hand, we conclude the same for L^p functions. The details are standard and omitted here. ∎

In the Euclidean case we have the sphere differentiation for L^p for all $p > \frac{n}{n-1}$. Likewise we expect that the above theorem is true for all $p > \frac{2n}{2n-1}$. We are falling short of the optimal result by half a derivative. We also note that the above theorem is valid only for $n \geq 2$. The case $n = 1$ plays a special role much as in the Euclidean case. Indeed, the maximal inequality for the spherical means on H^1 fails in L^2, and therefore, the method of square functions is not available in this case. In the next chapter we establish the optimal theorem for a restricted class of actions, namely those that factor to actions of the reduced Heisenberg group.

3.7   Notes and references

The Heisenberg motion group was studied by Strichartz [71]. The restriction theorem, Theorem 3.1.3 was proved in [54]. This theorem is very much in the spirit of a corresponding theorem for the spherical harmonic projections proved in Sogge [61]. For more about Gelfand pairs and spherical functions, we refer to Benson et al [7] and the references thereof. The Wiener-Tauberian theorem for $L^1(H^n/U(n))$ was proved in Hulanicki-Ricci [35] where the authors have applied the theorem to study tangential convergence of boundary harmonic functions on balls in $\mathbb{C}^n$. The version Theorem 3.4.2 is from Rawat [55]. Spherical means on the Heisenberg group have been studied by the author [87] and Agranovsky et al [1]. In both works a characterisation for CR functions is also given. The maximal theorem for spherical means is taken from Nevo-Thangavelu [50] where the authors have also proved pointwise ergodic theorems for the spherical means. A different type of spherical means has been studied by Cowling [4] using a different method.

Chapter 4

THE REDUCED HEISENBERG GROUP

Many problems on the Heisenberg group do not have satisfactory answers due to the fact that the centre is not compact. We have observed this in connection with the restriction theorem, the Wiener-Tauberian theorem and the maximal theorem for the spherical means. Some of these problems behave better when studied on the reduced Heisenberg group or Heisenberg group with compact centre. By taking the Fourier series of functions in the last variable, we can reduce problems on this group to problems on the phase space $\mathbb{C}^n$ equipped with the twisted convolution structure. In this chapter our aim is to study some of the problems treated in previous chapters in the setting of the reduced Heisenberg group.

4.1 The reduced Heisenberg group

In the Heisenberg group consider the central subgroup Γ given by

$$\Gamma = \{(0, 2\pi k) : k \in Z\}.$$

The reduced Heisenberg group is defined to be the quotient group H^n/Γ. The group operation is still given by

$$(z, t)(w, s) = (z + w, t + s + \frac{1}{2}Im(z.\bar{w}))$$

where now $(t + s + \frac{1}{2}Im(z.\bar{w}))$ is taken modulo 2π. The centre of the reduced Heisenberg group is then compact and can be identified with the interval $[0, 2\pi)$. An equivalent definition of the reduced Heisenberg group is $H^n/\Gamma = \mathbb{C}^n \times S^1$ with the group law

$$(z, e^{it})(w, e^{is}) = (z + w, e^{i(t+s+\frac{1}{2}Im(z.\bar{w}))}).$$

The one-dimensional representations χ_w of H^n are still representations of H^n/Γ but the Schrödinger representation π_λ will be a representation of H^n/Γ only in the case when λ is a nonzero integer. These are the relevant representations needed to build up the Plancherel theorem. Unlike the Heisenberg group, the Plancherel measure for H^n/Γ involves χ_w as well.

Recall the definition of e_k^j from Chapter 2. We can expand functions in $L^2(H^n/\Gamma)$ which have mean value $\int_0^{2\pi} f(z,t)\, dt = 0$ in terms of e_k^j. For such functions, the Plancherel theorem takes the following form.

Theorem 4.1.1 *Let $f \in L^2(H^n/\Gamma)$ be such that $\int_0^{2\pi} f(z,t)\, dt = 0$. Then we have*

$$\|f\|_2^2 = (2\pi)^{-2n-1} \sum_{k=0}^{\infty} \sum_{j=-\infty}^{\infty} j^{2n} \int_{\mathbb{C}^n} |f * e_k^j(z,0)|^2\, dz.$$

Proof: As $e_k^j(z,t) = e^{ijt}\varphi_k^j(z)$ we have

$$f * e_k^j(z,t) = e^{-ijt} f^j *_j \varphi_k^j(z).$$

Making a change of variables, we see that

$$f^j *_j \varphi_k^j(|j|^{-\frac{1}{2}}z) = |j|^{-n} f_j^j \times \varphi_k(z)$$

where $f_j^j(z) = f^j(|j|^{-\frac{1}{2}}z)$. Now,

$$\int_{\mathbb{C}^n} |f * e_k^j(z,0)|^2\, dz$$

$$= |j|^{-n} \int_{\mathbb{C}^n} |f^j *_j \varphi_k^j(|j|^{-\frac{1}{2}}z)|^2\, dz$$

and hence, in view of the above relation we have

$$\sum_{k=0}^{\infty} \int_{\mathbb{C}^n} |f * e_k^j(z,0)|^2\, dz$$

$$= |j|^{-3n} \sum_{k=0}^{\infty} \int_{\mathbb{C}^n} |f_j^j \times \varphi_k(z)|^2\, dz.$$

Appealing to the Plancherel theorem for the special Hermite expansions, we get

$$\sum_{k=0}^{\infty} \int_{\mathbb{C}^n} |f_j^j \times \varphi_k(z)|^2\, dz$$

$$= (2\pi)^{2n}|j|^n \int_{\mathbb{C}^n} |f^j(z)|^2 \, dz.$$

Finally,

$$\sum_{k=0}^{\infty} \sum_{j \neq 0} j^{2n} \int_{\mathbb{C}^n} |f * e_k^j(z,0)|^2 \, dz$$

$$= (2\pi)^{2n} \int_{\mathbb{C}^n} \left(\sum_{j \neq 0} |f^j(z)|^2 \right) dz.$$

An application of the Plancherel theorem for the Fourier series completes the proof. ∎

Given a function f on $L^2(H^n/\Gamma)$, we can write $f = g + h$ where

$$h(z) = (2\pi)^{-1} f^0(z)$$

and g has mean value zero. The Plancherel theorem can be written in the form

$$\|f\|_2^2 = \|h\|_2^2 + (2\pi)^{-2n-1}$$

$$\times \left(\sum_{k=0}^{\infty} \sum_{j=-\infty}^{\infty} j^{2n} \int_{\mathbb{C}^n} |g * e_k^j(z,0)|^2 \, dz \right).$$

Since $\|h\|_2$ can be expressed in terms of its Euclidean Fourier transform, which is the same as $\chi_w(h)$, we see that the Plancherel theorem involves both kinds of representations.

We now proceed to prove a restriction theorem for the spectral projections. Since the functions $f * e_k^j$ are eigenfunctions of the sublaplacian with eigenvalues $(2k+n)|j|$, it is natural to study the mapping properties of the operators

$$Q_N f = \sum_{(2k+n)|j|=N} f * e_k^j.$$

Unlike their counterparts on the full Heisenberg group, these operators are bounded from L^p into L^2. Let

$$d(N) = \sum_{(2k+n)|N} (2k+n)^{-1}$$

where $a|b$ means a divides b.

Theorem 4.1.2 *Let* $1 \leq p \leq 2$ *and* $f \in L^p(H^n/\Gamma)$*. Then*

$$\|Q_N f\|_2 \leq C(N^n d(N))^{(\frac{1}{p}-\frac{1}{2})}\|f\|_p.$$

When $p = 1$*, the above estimate is sharp.*

Proof: Since Q_N is bounded on L^2 with norm 1 it is enough to prove the result for $p = 1$. But then the result follows from Young's inequality. Indeed,

$$\|Q_N f\|_2 \leq \|f\|_1 \|\psi_N\|_2$$

where $\psi_N(z, t) = \sum_{(2k+n)|j|=N} e_k^j(z, t)$ and a calculation shows that

$$\|\psi_N\|_2^2 = c_n \sum_{(2k+n)|j|=N} |j|^n \|\varphi_k\|_2^2.$$

But $\|\varphi_k\|_2 = O((2k + n)^{n-1})$ and therefore,

$$\|\psi_N\|_2^2 = O(N^n d(N)).$$

This proves the result for $p = 1$. By taking an approximate identity f_ϵ, we know that $f_\epsilon * \psi_N$ converges to ψ_N in L^2 and so

$$\|\psi_N\|_2 \leq \limsup \|f_\epsilon * \psi_N\|_2 \leq \|Q_N\|_{1,2}$$

where $\|Q_N\|_{1,2}$ is the operator norm of Q_N from L^1 into L^2. The above proves the sharpness of the bound when $p = 1$. ∎

We can also get a lower bound for the operator norm $\|Q_N\|_{p,2}$ of Q_N taking L^p into L^2. To obtain this lower bound, let $N = nR$ where R is a positive integer and consider the function $f(z, t) = e^{-iRt}g(z)$ where g is a radial function on $\mathbb{C}^n$. Then it is clear that

$$Q_N f(z, t) = f * e_0^R(z, t)$$

which reduces to $R^n e^{-iRt} g *_R \varphi_0^R(z)$. A simple change of variables shows that

$$Q_N f(z, t) = e^{-iRt} G_R(\sqrt{R}z)$$

where

$$G_R(z) = \int_{\mathbb{C}^n} g(R^{-\frac{1}{2}}(z - w))e^{-\frac{1}{4}|w|^2}e^{-\frac{i}{2}Imz.\bar{w}} \, dw.$$

Since g is radial it follows that

$$G_R(z) = e^{-\frac{1}{4}|z|^2} \int_{\mathbb{C}^n} g(R^{-\frac{1}{2}}w)e^{-\frac{1}{4}|w|^2} \, dw.$$

We can choose g such that

$$\left| \int_{\mathbb{C}^n} g(R^{-\frac{1}{2}}w)e^{-\frac{1}{4}|w|^2} \, dw \right| \geq CR^{\frac{n}{p}} \|g\|_p$$

which will then imply that $\|G_R\|_2 \geq CR^{\frac{n}{p}}\|g\|_p$. Hence, we finally have

$$\|f * e_0^R\|_2 \geq CR^{n(\frac{1}{p}-\frac{1}{2})}\|g\|_p$$

which proves the lower bound as $\|f\|_p = \|g\|_p$.

4.2   A Wiener-Tauberian theorem for L^p functions

In this section we will establish a Wiener Tauberian theorem for L^p functions on the reduced Heisenberg group. Given $f \in L^p(H^n/\Gamma)$, we consider $V(f)$ which is the smallest closed subspace containing f that is invariant under the action of the Heisenberg motion group, i.e., invariant under translations and rotations. We are interested in finding conditions on f so that $V(f)$ is a proper subspace. Writing the Fourier series of $f(z,.)$, we are led to consider subspaces of $L^p(\mathbb{C}^n)$ that are invariant under rotations and twisted translations. So, we start with a study of such subspaces.

Given λ real, let $\tau_\lambda(z)$ be the λ-twisted translation defined by

$$\tau_\lambda(z)f(w) = f(w+z)e^{\frac{i}{2}\lambda Imw.\bar{z}}.$$

Note that when $\lambda = 0$, this is just the ordinary translation. For $\sigma \in U(n)$ let R_σ be the rotation operator defined by $R_\sigma f(w) = f(\sigma w)$. Given $f \in L^p(\mathbb{C}^n)$ let $V_\lambda(f)$ be the smallest closed subspace containing f which is invariant under $\tau_\lambda(z)$ and R_σ for all $z \in \mathbb{C}^n$ and $\sigma \in U(n)$. We first prove the following theorem.

Theorem 4.2.1 *Let $\lambda \neq 0$ and $f \in L^p(\mathbb{C}^n), 1 \leq p < \infty$. Then $V_\lambda(f)$ is proper if and only if $f *_\lambda \varphi_k^\lambda = 0$ for some k.*

Proof: We will prove the theorem when $\lambda = 1$. The general case is similar. If $V_1(f)$ is proper, then the subspace

$$V_1(f)^\perp = \{g \in L^{p'}(\mathbb{C}^n) : (h, \bar{g}) = 0, \forall h \in V_1(f)\}$$

is nontrivial. Let $g \in V_1(f)^\perp$ be a nontrivial function. Then it verifies $f \times g = 0$. We claim that this g can be taken to be radial. To see this, we first observe that as $V_1(f)$ is invariant under rotations, we also have $(R_\sigma h, \bar{g}) = 0$ for all σ and consequently the radialisation $g_0(z) = \int_K g(\sigma z)\, d\sigma$ also satisfies $(h, \bar{g}_0) = 0$ for all $h \in V_1(f)$. Our claim would be proved if we can show that for some $g \in V_1(f)^\perp$, its radialisation is nontrivial.

Suppose not. That is, $g_0 = 0$ for any $g \in V_1(f)^\perp$. We claim that this forces $V_1(f)$ to contain all radial functions in $L^p(\mathbb{C}^n)$. To see this, assume the contrary. If h is a radial function which is not in $V_1(f)$, we can find $\psi \in V_1(f)^\perp$ with $(h, \bar{\psi}) \neq 0$. As h is radial, the radialisation ψ_0 also satisfies $(h, \bar{\psi}_0) \neq 0$ and so ψ_0 is nontrivial. But then ψ_0 is a nontrivial radial function in $V_1(f)^\perp$ contradicting our assumption. So $V_1(f)$ contains all radial functions, in particular all φ_k. But then $V_1(f)^\perp$ cannot be nontrivial. This proves that there is a nontrivial radial g in $V_1(f)^\perp$.

For this radial g, $g \times \varphi_k = R_k(g)\varphi_k \neq 0$ for some k. But then $f \times g = 0$ gives

$$f \times g \times \varphi_k = R_k(g) f \times \varphi_k = 0.$$

As $R_k(g) \neq 0$ we necessarily have $f \times \varphi_k = 0$. This proves the direct half of the theorem. The converse is trivial: if $f \times \varphi_k = 0$ then $V_1(f)$ cannot contain φ_k.   ∎

We also have an analogue of this result for $\lambda = 0$. To prove such a result for $V_0(f)$, we need the following theorem which is Wiener's theorem for $\mathbb{R}^n$ in disguise. Given a function f on $\mathbb{R}^n$, we let $Z(f) = \{x : f(x) = 0\}$ stand for the zero set of f.

Theorem 4.2.2 *Let $f \in L^1(\mathbb{R}^n)$ and let $h \in L^\infty(\mathbb{R}^n)$ be such that $f * h = 0$. Then $supp(\hat{h}) \subset Z(\hat{f})$.*

Proof: Let x be any point in the complement of $Z(\hat{f})$. Without loss of generality we can assume that $\hat{f}(x) = 1$. We claim that there exists an L^1 function g with $\|g\|_1 < 1$ such that $\hat{g} = 1 - \hat{f}$ in some neighbourhood U of x. Assuming the claim for a moment, we will prove the theorem.

The theorem will follow if we can show that $\hat{h} = 0$ in U or equivalently $(\hat{h}, \hat{\psi}) = 0$ for every Schwartz class function ψ whose Fourier transform is supported in U. Since $(\hat{h}, \hat{\psi}) = h * \psi(0)$, it suffices to show that $h * \psi = 0$. Given such a function ψ, let us define $\psi_0 = \psi$ and $\psi_k = g * \psi_{k-1}$ for $k \geq 1$. Then

$$\|\psi_k\|_1 \leq \|g\|_1^k \|\psi\|_1$$

and since $\|g\|_1 < 1$, the function $G(x) = \sum \psi_k(x)$ is integrable. Moreover, as $\hat{g} = 1 - \hat{f}$ on the support of $\hat{\psi}$ we have

$$(1 - \hat{g}(y))\hat{\psi}(y) = \hat{\psi}(y)\hat{f}(y).$$

This means

$$\hat{\psi}(y) = \sum_{k=0}^{\infty} \hat{g}^k(y)\hat{\psi}(y)\hat{f}(y) = \hat{G}(y)\hat{f}(y).$$

Thus $\psi = G * f$ and so $\psi * h = G * f * h = 0$. This proves the theorem.

Coming to the proof of the claim, choose φ in $L^1(\mathbb{R}^n)$ such that $\hat{\varphi} = 1$ near the origin. For $r > 0$ let

$$\varphi_r(y) = e^{ix.y}r^{-n}\varphi(r^{-1}y)$$

and define

$$g_r(y) = \varphi_r(y) - f * \varphi_r(y).$$

Since $\hat{\varphi}_r = 1$ in some neighbourhood U_r of x, it follows that

$$\hat{g}_r(y) = 1 - \hat{f}(y)$$

for all $y \in U_r$. Now

$$g_r(y) = \int_{\mathbb{R}^n} f(u)(e^{-ix.u}\varphi_r(u) - \varphi_r(y - u))\, du.$$

Recalling the definition of φ_r, we get the estimate

$$\|g_r\|_1 \leq \left(\int_{\mathbb{R}^n} |f(y)|\, dy \right)$$

$$\times \left(\int_{\mathbb{R}^n} |\varphi(u) - \varphi(u - r^{-1}y)|\, du \right).$$

But the inner integral is at most $2\|\varphi\|_1$ and it tends to zero for every y as r tends to infinity. So we can choose r large enough so that $g = g_r$ will have norm strictly less than 1 and the claim is proved with $U = U_r$. ∎

Now we can prove the following result for $V_0(f)$. Recall that this is the smallest closed subspace invariant under ordinary translations and rotations which contains f.

Theorem 4.2.3 *Let $f \in L^1 \cap L^p(\mathbb{C}^n), 1 < p < \frac{4n}{2n+1}$. Then $V_0(f)$ is proper if and only if $f * \eta_\tau = 0$ for some $\tau > 0$.*

Proof: Note that

$$\eta_\tau(z) = c_n(\tau|z|)^{-n+1} J_{n-1}(\tau|z|).$$

It follows from the asymptotic properties of the Bessel function that $\eta_\tau \in L^q(\mathbb{C}^n)$ precisely when $q > \frac{4n}{2n-1}$ which is the conjugate index of $p = \frac{4n}{2n+1}$. If $f * \eta_\tau = 0$ for some $\tau > 0$ then clearly $V_0(f)$ cannot be the whole space $L^p(\mathbb{C}^n)$.

Observe that the condition $f * \eta_\tau = 0$ is equivalent to the fact that $\hat{f}$ vanishes on the sphere $|z| = \tau$. This follows from the definition of η_τ. Now to prove the converse, let $V_0(f)$ be proper and $\hat{f}$ does not vanish on any sphere $|z| = \tau$. Then there is $h \in L^{p'}(\mathbb{C}^n)$ which is orthogonal to all of $V_0(f)$. We can assume, without loss of generality, that h is radial and smooth. Convolving with a smooth compactly supported approximation identity, we can even assume that h is bounded. Thus $f * h = 0$. Since $\hat{f}$ may not be smooth, $\hat{f}\hat{h}$ need not be defined. However, by Theorem 4.2.2 we can conclude that $supp(\hat{h}) \subset Z(\hat{f})$.

Since $\hat{h}$ is radial, if $z \in supp(\hat{h})$ then

$$\{w : |w| = |z|\} \subset supp(\hat{h}) \subset Z(\hat{f}).$$

But $\hat{f}$ never vanishes on any sphere and so this is possible only for $z = 0$. This means that $supp(\hat{h}) = \{0\}$ and so h is a nontrivial polynomial which is impossible as $h \in L^{p'}(\mathbb{C}^n)$. This contradiction shows that $\hat{f}$ vanishes on some sphere. This completes the proof of the theorem.

Combining Theorems 4.2.1 and 4.2.2, we get the following result for the reduced Heisenberg group.

Theorem 4.2.4 *Let* $f \in L^1 \cap L^p(H^n/\Gamma)$, *and* $1 < p < \frac{4n}{2n+1}$. *Then* $V(f)$ *is proper if and only if one of the following conditions holds: (i)* $f * e_k^j = 0$ *for some* $j \neq 0$ *and* k. *(ii)* $f * \eta_\tau = 0$ *for some* $\tau > 0$.

Proof: As $V(f)$ is invariant under the action of the Heisenberg motion group, the functions $f((\sigma, z, t)(w, s))$ belong to $V(f)$ for all $(\sigma, z, t) \in G$ and so are their Fourier coefficients. A calculation shows that

$$\int_0^{2\pi} f((\sigma, z, t)(w, s)) e^{-ijt} \, dt$$

$$= e^{ijs} e^{\frac{i}{2}jIm\sigma w.\bar{z}} f^j(\sigma w + z)$$

where f^j is the j-th Fourier coefficient of $f(z,)$. This means that functions of the form

$$e^{ijs} e^{\frac{i}{2}jIm\sigma w.\bar{z}} f^j(\sigma w + z)$$

are all in $V(f)$ and consequently any function of the form $e^{ijs}h(z)$ with $h \in V_j(f^j)$ is in $V(f)$.

Now there is certainly a j for which $V_j(f^j)$ is proper. Otherwise, writing the Fourier series of any $g \in L^p(H^n/\Gamma)$ and using the above observations, we can conclude that $V(f) = L^p(H^n/\Gamma)$. So let us assume $V_j(f^j)$ is proper and $j \neq 0$. Then by Theorem 4.2.1, there is a k such that $f^j *_j \varphi_k^j = 0$. But then $f * e_k^j = 0$ as desired. If $j = 0$ we can apply Theorem 4.2.3 to conclude that $f * \eta_\tau = 0$ for some $\tau > 0$. This proves the theorem.

We conclude this section with the following two-radius theorem for tempered continuous functions on the reduced Heisenberg group. For a two-radius theorem for the spherical means on $\mathbb{R}^n$, see Delsarte [16].

Theorem 4.2.5 *Assume that f is a tempered continuous function on H^n/Γ which satisfies $f * \mu_r = f * \mu_s = 0$. Then $f = 0$ provided (i) $(\frac{r}{s})^2$ is not a quotient of zeros of $L_k^{n-1}(t)$ for any k (ii) $\frac{r}{s}$ is not a quotient of zeros of $J_{n-1}(t)$.*

Proof: We reduce the convolution equations $f * \mu_r = 0$ and $f * \mu_s = 0$ into twisted convolution equations by writing down their Fourier series. We have

$$f^j *_j \mu_r = f^j *_j \mu_s = 0$$

for any integer j. As in the proof of Theorem 3.5.6, we can assume that $f^j \in L^2(H^n/\Gamma)$ using the approximation theorems proved in Chapter 1. From the above equations we want to conclude that $f^j = 0$ for all j, which will prove the theorem.

Let $j > 0$ and expand $f^j *_j \mu_\rho$ in terms of $f^j *_j \varphi_k^j$. Using Proposition 3.5.5 we get

$$\sum_{k=0}^{\infty} \frac{k!(n-1)!}{(k+n-1)!} \varphi_k^j(\rho) f^j *_j \varphi_k^j(z) = 0$$

for $\rho = r, s$. Since for each k, either $\varphi_k^j(r) \neq 0$ or $\varphi_k^j(s) \neq 0$, we conclude that $f^j *_j \varphi_k^j = 0$ for all k. This means that $f^j = 0$. Similarly we conclude that $f^j = 0$ when j is negative. When $j = 0$, we have

$$\int_{|w|=\rho} f^0(z-w)\, d\mu_\rho = 0$$

for $\rho = r, s$. But then by the two-radius theorem for the ordinary convolution on $\mathbb{C}^n$ we get $f^0 = 0$. Thus all the Fourier coefficients of f are zero proving the result. ∎

4.3   A maximal theorem for spherical means

In this section we prove an optimal theorem for the maximal function
associated to the spherical means on the reduced Heisenberg group.
Here we consider the global part of the maximal function, namely we
consider

$$f_\mu^*(z,t) = \sup_{r \geq 1} |f * \mu_r(z,t)|.$$

As before, we use analytic interpolation arguments, but instead of Riemann-
Liouville fractional integrals, we use a different family defined in terms
of Laguerre functions L_k^α with α complex. Our estimates for the maximal function will be divided into two parts, as follows. Given a function
$f \in L^2(H^n/\Gamma)$, we can write $f = f_L + f_B$ where the spectral measure
of f_L is supported in Σ_L and that of f_B is supported in Σ_B. Since the
case of $f = f_B$ is the classical one studied by Stein in his theorem on
spherical means, we restrict our attention to $f = f_L$.

Recall that the Laguerre spectrum is given by

$$\Sigma_L = \{(\lambda, k) : \lambda \neq 0, k \in N\}.$$

As before, we write $\zeta = (\lambda, k)$ and $\varphi_\zeta(r)$ stands for $\varphi_\zeta(z, 0)$ with $|z| = r$.
Consider the Laguerre functions

$$\psi_k^\alpha(r) = \frac{\Gamma(k+1)\Gamma(\alpha+1)}{\Gamma(k+\alpha+1)} L_k^\alpha(\tfrac{1}{2}r^2) e^{-\frac{r^2}{4}}.$$

The above functions are defined even for complex values of α and we
have used them already in the proof of the restriction theorem for the
special Hermite projections. Using these operators, we define an analytic family operators by setting

$$(M_r^\alpha f, g) = \int_{\Sigma_L} \psi_k^{\alpha+n-1}(\sqrt{|\lambda|}r) \, d\nu_{f,g}(\lambda, k)$$

where, as before, $d\nu_{f,g}$ is the measure determined by f and g in the
spectral decomposition of the operator $\pi(m_r)$. Note that M_r^α is defined
as long as $Re(\alpha) > -n + 1$ and when $\alpha = 0$ we recover the spherical
means $f * \mu_r$. For the maximal function associated to M_r^α, we first prove
the following.

Proposition 4.3.1 *The   maximal   function $\sup_{r>0} |M_r^{1+ib} f|$ is
bounded on $L^p(H^n/\Gamma)$ for $1 < p < \infty$.*

We prove this proposition using the following representation of the operator M_r^1 in terms of the Poisson integrals. Let $P_r f$ stand for the ordinary Poisson integral of f in the central variable, defined by the equation

$$(P_r f)^j(z) = e^{-\frac{1}{4}|j|r} f^j(z).$$

We then have

Proposition 4.3.2 *As an operator on $L^2(H^n/\Gamma)$, M_r^1 is given by*

$$M_r^1 f(z, t) = \int_0^1 s^{2n-1} P_{r^2(1-s^2)} f * \mu_{rs}(z, t)\, ds.$$

Proof: The formula is proved using certain identities related to Laguerre functions. We write $f * \mu_r$ as

$$f * \mu_r(z, t) = \frac{1}{2\pi} \sum_{j=-\infty}^{\infty} e^{-ijt} (f * \mu_r)^j(z).$$

Now

$$(f * \mu_r)^j(z) = f^j *_j \mu_r(z)$$

and using the result of Proposition 3.5.5 we have

$$(f * \mu_r)^j(z)$$

$$= \sum_{k=0}^{\infty} \frac{k!(n-1)!}{(k+n-1)!} \varphi_k^j(r) f^j *_j \varphi_k^j(z).$$

Thus the spherical means $f * \mu_r$ has the following representation:

$$f * \mu_r(z, t)$$

$$= \frac{1}{2\pi} \sum_{k=0}^{\infty} \sum_{j=-\infty}^{\infty} e^{-ijt} \psi_k^{n-1}(\sqrt{|j|}r) f^j *_j \varphi_k^j(z).$$

Now we consider the equation

$$(2\pi) P_{r^2(1-s^2)} f * \mu_{rs}(z, t) =$$

$$\sum_{k=0}^{\infty} \sum_{j=-\infty}^{\infty} e^{-ijt} \psi_k^{n-1}(\sqrt{|j|}rs) e^{-\frac{1}{4}|j|r^2(1-s^2)} f^j *_j \varphi_k^j(z).$$

Integrating the above equation with respect to $s^{2n-1}ds$, we obtain

$$\int_0^1 s^{2n-1} P_{r^2(1-s^2)} f * \mu_{rs}(z,t)\, ds$$

$$= \frac{1}{2\pi} \sum_{k=0}^\infty \sum_{j=-\infty}^\infty e^{-ijt} \rho_k(\sqrt{|j|}r) f^j *_j \varphi_k^j(z)$$

where we have written

$$\rho_k(r) = \int_0^1 s^{2n-1} \psi_k^{n-1}(rs) e^{-\frac{1}{4}r^2(1-s^2)}\, ds.$$

Recalling the definition of ψ_k^{n-1}, we have

$$\rho_k(r) = \frac{\Gamma(k+1)\Gamma(n)}{\Gamma(k+n)}$$

$$\times \left(\int_0^1 s^{2n-1} L_k^{n-1}(\frac{1}{2}r^2 s^2) e^{-\frac{1}{4}r^2}\, ds \right).$$

We now use the following formula (see formula 7.4.11 in [84]) connecting Laguerre polynomials of different type:

$$L_k^{\alpha+\beta}(r) = \frac{\Gamma(k+\alpha+\beta+1)}{\Gamma(\beta)\Gamma(k+\alpha+1)}$$

$$\times \left(\int_0^1 s^\alpha (1-s)^{\beta-1} L_k^\alpha(sr)\, ds \right).$$

Taking $\alpha = n-1$ and $\beta = 1$ in this formula, we see that

$$\rho_k(r) = \frac{1}{2n} \psi_k^n(r).$$

Thus we have the formula

$$M_r^1 f(z,t) = \int_0^1 s^{2n-1} P_{r^2(1-s^2)} f * \mu_{rs}(z,t)\, ds$$

and the proposition is proved.   ∎

Now we can prove Proposition 4.3.1. Changing variables, we have

$$M_r^1 f(z,t) = r^{-2n} \int_0^r s^{2n-1} P_{(r^2-s^2)} f * \mu_s(z,t)\, ds.$$

Recall the well known fact that

$$\sup_{r>0} |P_r f(z,t)| \leq C\Lambda_1 f(z,t)$$

where $\Lambda_1 f$ is the Hardy-Littlewood maximal function in the t variable. Using this we see that

$$|M_r^1 f(z,t)| \leq Cr^{-2n} \int_0^r s^{2n-1} \Lambda_1 f * \mu_s(z,t) \, ds$$

which in turn is dominated by

$$Cr^{-1} \int_0^r \Lambda_1 f * \mu_s(z,t) \, ds.$$

As in the case of the Heisenberg group, the maximal function associated to the above uniform averages is bounded on L^p. Since Λ_1 is also bounded on L^p, the proposition is proved. ∎

In order to apply the method of analytic interpolation, we need an L^2 estimate for negative values of α, which is provided by the following proposition.

Proposition 4.3.3 *For a function $f \in L^2(H^n/\Gamma)$ whose spectral measure has its support in Σ_L the estimate*

$$\| \sup_{r \geq 1} |M_r^\alpha f| \|_2 \leq C\|f\|_2$$

holds whenever $\alpha = -n + 1 + \delta$ where $\delta > 0$.

Proof: We divide the Laguerre spectrum into two parts, as follows. For $j = 0, 1, 2, \ldots$, define

$$\Sigma_j = \{(\lambda, k) : |\lambda| k \in [2^j, 2^{j+1})\}$$

and

$$\Sigma^0 = \cup_{j \geq 0} \Sigma_j = \{(\lambda, k) : |\lambda| k \geq 1\}$$

Let Pf be the orthogonal projection corresponding to the set Σ^0. Note that

$$\Sigma_L - \Sigma^0 = \{(\lambda, 0) : \lambda \neq 0\}.$$

We will begin by establishing spectral estimates on Σ^0 first, and comment on its complement in Σ_L later.

Let $E_j f$ be the projection operator corresponding to the set Σ_j. It suffices to show that

$$\| \sup_{r \geq 1} |M_r^\alpha E_j f| \|_2 \leq C 2^{-\frac{1}{2}\delta j} \|f\|_2.$$

Now we use the elementary inequality

$$\sup_{r \geq 1} |g(r)|^2 \leq C\{L^{-1} \int_1^\infty |g(t)|^2\, dt + L \int_1^\infty |g'(t)|^2\, dt\}$$

for every $L > 0$. This follows from Cauchy-Schwarz and the inequality

$$2ab \leq La^2 + L^{-1}b^2$$

(see Stein [65] for a proof). Applying the above inequality, we have

$$\sup_{r \geq 1} |M_r^\alpha f(z,t)|^2$$

$$\leq C\{L^{-1} \int_1^\infty |M_r^\alpha f(z,t)|^2\, dr + L \int_1^\infty |\frac{d}{dr} M_r^\alpha f(z,t)|^2\, dr.\}$$

Now, as $\alpha = -n + 1 + \delta$, we have

$$(M_r^\alpha E_j f, g) = \int_{\Sigma_j} \psi_k^\delta(\sqrt{|\lambda|}r)\, d\nu_{f,g}$$

and therefore, in view of Proposition 3.3.6 and the above inequality, we need to show that

$$L^{-1} \int_1^\infty |\psi_k^\delta(\sqrt{|\lambda|}r)|^2\, dr + L \int_1^\infty |\frac{d}{dr}\psi_k^\delta(\sqrt{|\lambda|}r)|^2\, dr$$

is bounded by $C 2^{-\delta j}$. Assuming $\lambda > 0$ and changing variables, we have

$$\int_1^\infty |\psi_k^\delta(\sqrt{|\lambda|}r)|^2\, dr = \lambda^{-\frac{1}{2}} \int_{\sqrt{\lambda}}^\infty |\psi_k^\delta(r)|^2\, dr$$

and

$$\int_1^\infty |\frac{d}{dr}\psi_k^\delta(\sqrt{|\lambda|}r)|^2\, dr = \lambda^{\frac{1}{2}} \int_{\sqrt{\lambda}}^\infty |\frac{d}{dr}\psi_k^\delta(r)|^2\, dr.$$

Recalling the definition of the normalised Laguerre functions $\mathcal{L}_k^\alpha$ and noting the estimate

$$\frac{\Gamma(k+1)\Gamma(\delta+1)}{\Gamma(k+\delta+1)} = O(k^{-\delta})$$

and the formula $\frac{d}{dr}L_k^\delta(r) = -L_{k-1}^{\delta+1}(r)$, we are led to the estimation of J_1 and J_2 where

$$J_1 = \lambda^{-\frac{1}{2}}k^{-\delta}\int_\lambda^\infty |\mathcal{L}_k^\delta(r)|^2 r^{-\delta-\frac{1}{2}}\,dr$$

and

$$J_2 = \lambda^{\frac{1}{2}}k^{-\delta+1}\int_\lambda^\infty |\mathcal{L}_{k-1}^{\delta+1}(r)|^2 r^{-\delta-\frac{1}{2}}\,dr+$$

$$+\lambda^{\frac{1}{2}}k^{-\delta}\int_\lambda^\infty |\mathcal{L}_k^\delta(r)|^2 r^{-\delta+\frac{1}{2}}\,dr.$$

To estimate the above integrals, we need to use the following asymptotic properties of the Laguerre functions (see Szegö [72] and Lemma 1.5.3 of [84]).

Lemma 4.3.4 *Let $\alpha \geq 0$ and $k \geq 1$. Then we have the following estimates for the Laguerre functions: (i) $|\mathcal{L}_k^\alpha(r)| \leq C(kr)^\alpha$, when $0 \leq r \leq \frac{1}{k}$ (ii) $|\mathcal{L}_k^\alpha(r)| \leq C(kr)^{-\frac{1}{4}}$, when $\frac{1}{k} \leq r \leq \frac{k}{2}$ (iii) $|\mathcal{L}_k^\alpha(r)| \leq Ck^{-\frac{1}{4}}(1 + |k - r|)^{-\frac{1}{4}}$, when $\frac{k}{2} \leq r \leq \frac{3k}{2}$ and (iv) $|\mathcal{L}_k^\alpha(r)| \leq Ce^{-\gamma r}$, when $r \geq \frac{3k}{2}$.*

We now estimate the expression J_1 above. Noting that $\lambda \geq \frac{1}{k}$, we write J_1 as a sum of three integrals:

$$J_1 = \lambda^{-\frac{1}{2}}k^{-\delta}$$

$$\times\left(\int_\lambda^{\frac{k}{2}} + \int_{\frac{k}{2}}^{\frac{3k}{2}} + \int_{\frac{3k}{2}}^\infty |\mathcal{L}_k^\delta(r)|^2 r^{-\delta-\frac{1}{2}}\,dr\right).$$

Using the estimates of Lemma 4.3.4 we obtain

$$\int_\lambda^{\frac{k}{2}} |\mathcal{L}_k^\delta(r)|^2 r^{-\delta-\frac{1}{2}}\,dr$$

$$\leq Ck^{-\frac{1}{2}}\int_\lambda^{\frac{k}{2}} r^{-\delta-1}\,dr \leq k^{-\frac{1}{2}}\lambda^{-\delta}$$

and

$$\int_{\frac{k}{2}}^{\frac{3k}{2}} |\mathcal{L}_k^\delta(r)|^2 r^{-\delta-\frac{1}{2}}\,dr \leq Ck^{-\frac{1}{2}}$$

$$\times\left(\int_{\frac{k}{2}}^{\frac{3k}{2}} (1 + |k - r|)^{-\frac{1}{2}} r^{-\delta-\frac{1}{2}}\,dr\right)$$

which is bounded by $Ck^{-\frac{1}{2}}\lambda^{-\delta}$ as $\lambda \leq k$. Similarly, the third integral is bounded by $Ck^{-\frac{1}{2}}\lambda^{-\delta}$ and so we have the estimate $J_1 \leq C(\lambda k)^{-\delta-\frac{1}{2}}$.

To estimate J_2 we consider the first integral appearing in the expression for J_2 which is a sum of three integrals:

$$\lambda^{\frac{1}{2}}k^{-\delta+1}\left(\int_{\lambda}^{\frac{k}{2}} + \int_{\frac{k}{2}}^{\frac{3k}{2}} + \int_{\frac{3k}{2}}^{\infty} |\mathcal{L}_k^{\delta+1}(r)|^2 r^{-\delta-\frac{1}{2}}\, dr\right).$$

Using the estimates of the lemma again we get

$$\int_{\lambda}^{\frac{k}{2}} |\mathcal{L}_k^{\delta+1}(r)|^2 r^{-\delta-\frac{1}{2}}\, dr$$

$$\leq Ck^{-\frac{1}{2}}\int_{\lambda}^{\frac{k}{2}} r^{-\delta-1}\, dr \leq Ck^{-\frac{1}{2}}\lambda^{-\delta}.$$

Similary we can show that the other two parts also give the estimate $Ck^{-\frac{1}{2}}\lambda^{-\delta}$. The same is true of the second integral appearing in the expression for J_2. Putting all the estimates together, we have $J_2 \leq C(\lambda k)^{-\delta+\frac{1}{2}}$.

Finally, it follows that

$$\int_{H^n/\Gamma} \sup_{r\geq 1} |M_r^{\alpha} E_j f(h)|^2\, dh$$

$$\leq C\int_{\Sigma_j}\left(L^{-1}a(k,\lambda) + Lb(k,\lambda)\right)\, d\nu_f$$

where

$$a(k,\lambda) = \int_1^{\infty} |\psi_k^{\delta}(\sqrt{|\lambda|}r)|^2\, dr$$

and

$$b(k,\lambda) = \int_1^{\infty} |\frac{d}{dr}\psi_k^{\delta}(\sqrt{|\lambda|}r)|^2\, dr.$$

According to the foregoing estimates we have

$$L^{-1}a(k,\lambda) + Lb(k,\lambda)$$

$$\leq C\left(L^{-1}(\lambda k)^{-\delta-\frac{1}{2}} + L(\lambda k)^{-\delta+\frac{1}{2}}\right).$$

Recall that on Σ_j, $2^j \leq \lambda k \leq 2^{j+1}$ and therefore, choosing $L = 2^{-\frac{j}{2}}$ we get

$$L^{-1}a(k,\lambda) + Lb(k,\lambda) \leq C2^{-j\delta}.$$

Hence we have proved

$$\int_{H^n/\Gamma} \sup_{r \geq 1} |M_r^\alpha E_j f(h)|^2 \, dh$$

$$\leq C 2^{-j\delta} \int_{\Sigma_j} d\nu_f \leq C 2^{-j\delta} \|E_j f\|_2^2.$$

We note that in the case of the reduced Heisenberg group, the Laguerre part of the Gelfand spectrum is

$$\Sigma_L = \{(j, k) : j \neq 0, k = 0, 1, 2, \ldots\}.$$

Therefore, we have taken care of functions whose spectral measure is supported in Σ^0. The complement of Σ^0 is the set $\{(j, 0) : j \neq 0\}$ and in this case the estimates are easier as

$$\psi_0^\delta(\sqrt{|j|}r) = e^{-\frac{|j|}{4} r^2}.$$

Therefore, we obtain the proposition. ∎

In the above proposition we can take α to be complex with $Re(\alpha) = -n+1+\delta$ and get the same estimate. We can now interpolate the results of Propositions 4.3.1 and 4.3.2 to obtain the following theorem.

Theorem 4.3.5 *Let $n \geq 2$ and $\frac{2n}{2n-1} < p < \infty$. Then the global part of the maximal function f_μ^* associated to the spherical means satisfies the strong maximal inequality $\|f_\mu^*\|_p \leq C\|f\|_p$ for all $f \in L^p(H^n/\Gamma)$.*

4.4 Mean periodic functions on phase space

A continuous function f on $\mathbb{R}^n$ is said to be mean-periodic if the closed subspace $T(f)$ generated by f and all its translates is proper in $C(\mathbb{R}^n)$, the space of continuous functions with the topology of uniform convergence on compact sets. The fundamental theorem of mean-periodic functions, due to L. Schwartz [59], says that if f is mean-periodic on $\mathbb{R}$, then $T(f)$ contains an exponential function $e^{i\lambda x}$ for some $\lambda \in \mathbb{C}$. An exact analogue of this fails in the case of $\mathbb{R}^n, n \geq 2$ (see [32]). However, a weaker version of Schwartz theorem is true in many situations, including $\mathbb{R}^n$.

Instead of considering translations alone, we can also consider rotations. Let $V(f)$ be the smallest closed subspace of $C(\mathbb{R}^n)$ invariant under translations and rotations containing f. In [10], Brown et al proved that such a subspace contains the Bessel function

$$\varphi_\lambda(x) = (\lambda|x|)^{-\frac{n}{2}+1} J_{\frac{n}{2}-1}(\lambda|x|)$$

for some $\lambda \in \mathbb{C}$. Note that φ_λ are the elementary spherical functions on the Euclidean spaces. A similar result for noncompact symmetric spaces was established by Bagchi and Sitaram [3] and the case of the motion group was treated by Weit [91]. In all these cases it was proved that the appropriate subspace V contains an elementary spherical function.

Our aim in this section is to study mean-periodic functions on the Heisenberg group. If f is a mean-periodic function on H^n and if $V(f)$ is the translation and rotation invariant subspace of $C(H^n)$ generated by f, then we may ask if $V(f)$ contains an elementary spherical function. The result of Agranovsky et al concerning spherical means for bounded continuous functions (see Corollary 3.5.4) means that when f is a bounded mean-periodic function then $V(f)$ does contain an elementary spherical function. For the reduced Heisenberg group, we will show that a similar result is true for any mean periodic function of tempered growth. For the general case we make a conjecture which is yet to be settled.

Given a continuous function f on H^n, we say that it is mean-periodic if $T(f)$ is a proper subspace of $C(H^n)$. Here $C(H^n)$ is equipped with the topology of uniform convergence on compact subsets. When the subspace $V(f)$ is proper, we say that f is spherically mean-periodic. The study of mean-periodic functions on H^n is closely related to the study of twisted mean-periodic functions on the phase space $\mathbb{C}^n$ equipped with the symplectic form $[z, w] = Im(z.\bar{w})$. We say that a function f on the phase space is twisted mean-periodic if the closed subspace generated by f and all its twisted translations $\tau(w)f(z) = f(z + w)e^{\frac{i}{2}[z,w]}$ is proper. We can similarly define twisted spherically mean-periodic functions. In what follows we will omit the adjective twisted with the understanding that, whenever we talk about phase space, we always consider the twisted convolution structure.

If f is a mean periodic function of the phase space, then by the Hahn-Banach theorem, there exists a compactly supported Radon measure μ such that $f \times \mu = 0$. In the case of spherically mean-periodic functions, we can choose μ to be radial. We can take this property as an equivalent definition of mean-periodic functions. The simplest example of a mean-periodic function on phase space is given by the Laguerre functions. As we know

$$\varphi_k \times \mu_r = \frac{k!(n-1)!}{(k+n-1)!} \varphi_k(r) \varphi_k$$

and consequently $\varphi_k \times \mu_r = 0$ when r is a zero of $\varphi_k(t)$. Thus, φ_k is a (spherically) mean-periodic function. Similarly, the elementary spherical functions e_k^λ are all mean-periodic on the Heisenberg group.

We observe that the mean-periodic functions φ_k are Schwartz class functions. This is in sharp contrast with the ordinary mean-periodic functions on $\mathbb{R}^n$. As can be easily seen, no mean-periodic function on $\mathbb{R}^n$ can be integrable. Thus, though the study of integrable mean-periodic functions on $\mathbb{R}^n$ doesn't make sense, we can study such functions on the phase space. Our first result is the following theorem which is the analogue of the Schwartz theorem for integrable mean-periodic functions on $\mathbb{C}^n$. Recall that $V_\varphi(\psi)$ is the Fourier-Wigner transform of φ and ψ.

Theorem 4.4.1 *Let $f \in L^p(\mathbb{C}^n), 1 \leq p \leq 2$ be mean-periodic. Then there exists φ in $L^2(\mathbb{R}^n)$ such that $\overline{V_\varphi(\varphi)} \in T(f)$.*

Proof: We remark that $W(f)$ is well defined as a bounded operator on $L^2(\mathbb{R}^n)$. This follows from the fact that the Fourier-Wigner transform of two square integrable functions is in $L^p(\mathbb{C}^n)$ for $p \geq 2$ (see Corollary 1.2.3). As f is mean periodic, there is a nontrivial compactly supported Radon measure μ on $\mathbb{C}^n$ such that $f \times \mu = 0$. If $g \in L_0^2(\mathbb{C}^n)$, the subspace of compactly supported functions in $L^2(\mathbb{C}^n)$, it follows that $\mu \times g \in L_0^2(\mathbb{C}^n)$ and $f \times (\mu \times g) = 0$. So we let

$$T(f)^\perp = \{g \in L_0^2(\mathbb{C}^n) : f \times g = 0\}.$$

Let $f^*(z) = \bar{f}(-z)$ so that $W(f^*) = W(f)^*$. Therefore, if $g \in T(f)^\perp$ then $W(g^*)W(f^*) = 0$. Since $W(f^*)$ is a nontrivial bounded operator on $L^2(\mathbb{R}^n)$, there is $\psi \in L^2(\mathbb{R}^n)$ such that $\varphi = W(f^*)\psi$ is nontrivial. We will show that $\overline{V_\varphi(\varphi)} \in T(f)$.

We first observe that $W(g^*)\varphi = 0$ for all $g \in T(f)^\perp$. Noting the definition of $W(g^*)$, we have

$$0 = (W(g^*)\varphi, \varphi) = \int_{\mathbb{C}^n} \bar{g}(-z)(\pi(z)\varphi, \varphi)\, dz$$

which means that $\bar{g} \times V_\varphi(\varphi, 0) = 0$. Since $\tau(w)g$ also belongs to $T(f)^\perp$ whenever g does, we have $\bar{g} \times V_\varphi(\varphi, z) = 0$ for all z. This means that $\overline{V_\varphi(\varphi)} \times g = 0$ for all $g \in T(f)^\perp$ and so $\overline{V_\varphi(\varphi)} \in T(f)$. ∎

For the above proof, the fact that $W(f)$ is bounded on $L^2(\mathbb{R}^n)$ is crucial. When f is just a tempered distribution, we can still define $W(f)$ but that may not be a bounded operator on $L^2(\mathbb{R}^n)$. One can show that $W(f)$ will be a bounded operator between certain Hermite-Sobolev spaces, (see [88]). We do not need this result here. What we need is the fact that $(f, \Phi_{\alpha,\beta}) = 0$ for all α and β implies that $f = 0$. This follows from the fact that finite linear combinations of $\Phi_{\alpha,\beta}$ are dense in the Schwartz space (see Theorem 1.4.4). For spherically mean-periodic functions of tempered growth, we have the following result.

Theorem 4.4.2 *Let f be a continuous function of tempered growth on $\mathbb{C}^n$. If f is spherically mean-periodic, then $\varphi_k \in V(f)$ for some k. Consequently, $\Phi_{\alpha,\beta} \in V(f)$ for all α and β with $|\beta| = k$.*

Proof: Let $V(f)^{\perp} = \{g \in L_0^2(\mathbb{C}^n) : f \times g = 0\}$. We first claim that there is k such that $R_k(g) = 0$ for all radial functions g in $V(f)^{\perp}$. Here $R_k(g)$ is the k-th Laguerre coefficient of g. If the claim is not true, then for every k, we can find a radial $g \in V(f)^{\perp}$ such that $R_k(g) \neq 0$. But then, $f \times g = 0$ gives $f \times \varphi_k = 0$ for all k, which forces f to be zero. Hence the claim.

Now let k be such that $R_k(g) = 0$ for all radial functions g in $V(f)^{\perp}$. We will show that $\varphi_k \in V(f)$. First observe that if μ is a radial measure such that $f \times \mu = 0$, then $\mu(\varphi_k) = 0$. To see this, let g_n be any compactly supported approximate identity. Then $\mu \times g_n$ converges to μ weakly. As $\varphi_k \times \mu \times g_n = 0$ for all n, we get $\varphi_k \times \mu = 0$. Now, if φ_k is not in $V(f)$, then we can find a compactly supported Radon measure μ such that $f \times \mu = 0$ but $\mu(\varphi_k) \neq 0$. But then its nontrivial radialisation ν will satisfy $f \times \nu = 0$ and $\nu(\varphi_k) \neq 0$. This contradiction proves that $\varphi_k \in V(f)$.

Finally, $\varphi_k \in V(f)$ implies that $\varphi_k \times g = 0$ for all $g \in V(f)^{\perp}$. Consequently, $\bar{g} \times \varphi_k = 0$ or taking the Weyl transform, $W(\bar{g})P_k = 0$ where P_k is the k-th Hermite projection. But then, for any β with $|\beta| = k$ and $\alpha \in \mathbb{N}^n$, we have $(W(\bar{g})\Phi_{\beta}, \Phi_{\alpha}) = 0$ which is the same as

$$\int_{\mathbb{C}^n} \bar{g}(z)\Phi_{\beta,\alpha}(z)\,dz = 0.$$

Since $\Phi_{\alpha,\beta}(z) = \overline{\Phi_{\beta,\alpha}(-z)}$ we have $(g^*, \Phi_{\alpha,\beta}) = 0$ for all $g \in V(f)^{\perp}$. This proves that $\Phi_{\alpha,\beta} \in V(f)$. ∎

Using Theorems 4.4.1 and 4.4.2, we now prove the following result concerning mean-periodic functions on the reduced Heisenberg group.

Theorem 4.4.3 *(i) If f is an integrable mean-periodic function on H^n/Γ, then $T(f)$ contains either a function of the form $(\varphi, \pi_j(z,t)\varphi)$ for some $j \neq 0$ and $\varphi \in L^2(\mathbb{R}^n)$ or all functions $\eta(z)$ independent of t. (ii) If f is a spherically mean-periodic function of tempered growth on H^n/Γ then $V(f)$ contains either the Bessel function $\eta_\tau(z)$ for some $\tau \in \mathbb{C}$ or the function $e_k^j(z,t)$ for some k and $j \neq 0$.*

Proof: First consider the case of integrable mean-periodic functions. As f is nontrivial, for some j the function

$$f^j(z) = \int_0^{2\pi} f(z,t)e^{-ijt}\,dt$$

is nontrivial. Let $\tau_j(w)$ be the j−twisted translations defined earlier in this chapter. Theorems 4.4.1 and 4.4.2 have analogues for subspaces invariant under rotations and j-twisted translations. We observe that

$$e^{ijs}f^j(z+w)e^{\frac{i}{2}j[z,w]}$$

$$= \int_0^{2\pi} f((z,t)(w,s))e^{-ijt}\,dt$$

which shows that functions of the form

$$e^{ijs}f^j(z+w)e^{\frac{i}{2}j[z,w]}$$

are in $T(f)$.

First assume that $j \neq 0$ and let $T_j(f^j)$ be the closed subspace generated by f^j and its j-twisted translations. Then by Theorem 4.4.1, there is $\varphi \in L^2(\mathbb{R}^n)$ such that $(\varphi, \pi_j(z)\varphi) \in T_j(f^j)$. It then follows that $e^{ijt}(\varphi, \pi_j(z)\varphi) \in T(f)$. If $j = 0$, then $g \in T(f)$ whenever g belongs to the ordinary translation invariant subspace generated by f^0. As f^0 is nontrivial and integrable, this subspace is the whole of $C(\mathbb{C}^n)$, and hence $T(f)$ contains all functions independent of t.

The proof of (ii) is similar. If $j \neq 0$, then by Theorem 4.4.2 we conclude that $V(f)$ contains e_k^j for some k. When $j = 0$, we consider the ordinary translation and rotation invariant subspace generated by f^0. By the results of Brown et al [10] we conclude that $V(f)$ contains a Bessel function. We refer to [10] for details. ■

We now proceed to study mean-periodic functions of arbitrary growth. The fundamental theorem of mean-periodic functions on the real line is proved in the following way. Given a mean-periodic function

f on $\mathbb{R}$, we look at all compactly supported Radon measures μ such that $f * \mu = 0$ and consider J, the closed ideal generated by $\{\hat{\mu}(\zeta) : f * \mu = 0\}$ in $\hat{E}(\mathbb{R})$, the space of entire functions of exponential type. Then J is nontrivial whenever f is and the fundamental result on such ideals says that J has a common zero. This means that there is a $\zeta \in \mathbb{C}$ such that $\hat{\mu}(\zeta) = 0$ for all μ with $f * \mu = 0$. But then it is immediate that $e^{i\zeta x} \in T(f)$.

The above fundamental theorem on closed ideals in $\hat{E}(\mathbb{R}^n)$ is true only when $n = 1$. As was shown by Gurevich [32], there exist six distributions μ_j on $\mathbb{R}^2$ with compact support such that the ideal generated by $\{\hat{\mu}_j : j = 1, 2, \ldots, 6\}$ does not have a common zero. On the other hand, positive results are known in the case of ideals that are invariant under rotations. In [10] it was proved that any such proper ideal in $\hat{E}(\mathbb{R}^n)$ has a common zero. We like to formulate similar results for the Heisenberg group. Using the Fourier-Weyl transform, we associate closed ideals to any mean-periodic function on phase space, and ask for properties of those ideals that characterise mean-periodic functions.

Let f be a mean-periodic function on the phase space, and let $T(f)$ and $T(f)^\perp$ be defined as earlier. For $g \in T(f)^\perp$ let $\tilde{g}$ be its Fourier-Weyl transform. Define

$$I(f) = \{\tilde{g} : g^* \in T(f)^\perp\}.$$

Regarding $I(f)$ we have the following result. Recall the definition of E_0 from Section 1.5. It is the image of $L_0^2(\mathbb{C}^n)$ under the Fourier-Weyl transform. The elements of E_0 are entire functions of exponential type on $\mathbb{C}^{2n}$ taking values in the space of Hilbert-Schmidt operators and satisfying certain identity when restricted to $\mathbb{R}^{2n}$.

Proposition 4.4.4 *Let f be mean-periodic. Then $I(f)$ is a closed proper left ideal of E_0.*

Proof: To show that $I(f)$ is an ideal, let $\tilde{g} \in I(f)$ and $\tilde{h} \in E_0$. Then,

$$\tilde{h}(\zeta)\tilde{g}(\zeta) = (h \times g)\widetilde{\ }(\zeta)$$

which follows from the definition of the Fourier-Weyl transform and

$$(h \times g)^* = g^* \times h^*$$

shows that $(h \times g)^* \in T(f)^\perp$. Hence $\tilde{h}\tilde{g} \in I(f)$.

To show that $I(f)$ is closed in E_0, assume that $\tilde{g}_n \in T(f)^\perp$ converges to $\tilde{g} \in E_0$. This implies that $\tilde{g}_n(0)$ converges to $\tilde{g}(0)$ in $\mathcal{S}_2$, which means that $\|g_n - g\|_2 \to 0$. Moreover, all g_n and g are supported in a fixed compact set, say $|z| \le B$. Thus we have

$$f \times g^*(z) = f \times (g^* - g_n^*)(z)$$

$$= \int_{|w| \le B} f(z - w)(g^*(w) - g_n^*(w))e^{\frac{i}{2}[z,w]} \, dw$$

and an application of Cauchy-Schwarz gives

$$|f \times g^*(z)|^2$$

$$\le \|g_n - g\|_2^2 \int_{|w| \le B} |f(z - w)|^2 \, dw$$

which converges to zero as $n \to \infty$. This proves that $f \times g^* = 0$, and hence $g^* \in T(f)^\perp$.

Thus we have proved that $I(f)$ is a closed ideal. This cannot be the whole of E_0, because, otherwise $f \times g = 0$ for all $L_0^2(\mathbb{C}^n)$, which will force f to be zero by an approximate identity argument. $\blacksquare$

Our next proposition shows that the left ideal associated to a spherically mean-periodic function has a nice invariance property. Let f be spherically mean-periodic and let $V(f)^\perp$ be defined as above. Let

$$J(f) = \{\tilde{g} : g^* \in V(f)^\perp\}$$

be the associated left ideal. Using the metaplectic representation, we define an operator on E_0. Let $Sp(n)$ stand for the symplectic group consisting of $2n \times 2n$ matrices preserving the symplectic form

$$[z, w] = [(x, y), (u, v)] = y.u - x.v.$$

on $\mathbb{C}^n = \mathbb{R}^{2n}$. Given $\sigma = a + ib$ in $U(n)$ with a and b real matrices, the matrix $\begin{pmatrix} a & -b \\ b & a \end{pmatrix}$ is in $Sp(n)$ and for $\xi \in \mathbb{R}^{2n}$ we write $\sigma\xi$ to stand for $\begin{pmatrix} a & -b \\ b & a \end{pmatrix} \xi$. With this notation any $\sigma \in U(n)$ preserves the symplectic form: $[\sigma z, \sigma w] = [z, w]$. When $\zeta \in \mathbb{C}^{2n}$, $\sigma\zeta$ will stand for the element of $\mathbb{C}^{2n}$ obtained by applying the matrix $\begin{pmatrix} a & -b \\ b & a \end{pmatrix}$ to the $2n-$ vector ζ. For $\sigma \in U(n)$ we define an operator T_σ on E_0 by setting

$$T_\sigma \tilde{g}(\zeta) = \mu(\sigma)^* \tilde{g}(\sigma\zeta)\mu(\sigma)$$

where $\mu(\sigma)$ is the metaplectic representation.

Proposition 4.4.5 *The ideal $J(f)$ is invariant under the action T_σ for all $\sigma \in U(n)$.*

Proof: The invariance of $V(f)$ under the action of $U(n)$ shows that $g_\sigma^* \in V(f)^\perp$ whenever $g^* \in V(f)^\perp$ where $g_\sigma^*(z) = g^*(\sigma z)$. The proposition will follow once we establish $T_\sigma \tilde{g}(\zeta) = \tilde{g}_\sigma(\zeta)$. To see this, consider the Fourier-Weyl transform

$$\tilde{g}_\sigma(\zeta) = \int_{\mathbb{C}^n} e^{-i[z,\sigma\zeta]} g(z)\pi(z)\, dz.$$

By the remarks above, it follows that $[z,\zeta] = [\sigma z, \sigma\zeta]$ even for $\zeta \in \mathbb{C}^{2n}$. Therefore, we have

$$\tilde{g}_\sigma(\zeta) = \int_{\mathbb{C}^n} e^{-i[z,\zeta]} g(\sigma z)\pi(\sigma z)\, dz.$$

From the definition of the metaplectic representation it follows that

$$\tilde{g}(\sigma\zeta) = \mu(\sigma)\tilde{g}_\sigma(\zeta)\mu(\sigma)^*.$$

This proves the proposition. ∎

Having associated a closed left ideal to each spherically mean-periodic function, we may now ask whether the ideal has a common zero. The following proposition answers this question in the negative.

Proposition 4.4.6 *There is a spherically mean periodic function and a compactly supported Radon measure such that $f \times \mu = 0$ but $\tilde{\mu}(\zeta)$ does not vanish for any ζ.*

Proof: The example is provided by the Laguerre function φ_k. As we have already seen $\varphi_k \times \mu_r = 0$ for a proper choice of r. We claim that $\tilde{\mu}_r(\zeta) \neq 0$ for all $\zeta \in \mathbb{C}^{2n}$. For simplicity of notation, we assume $n = 1$. Suppose $\tilde{\mu}_r(\zeta) = 0$ for some $\zeta \in \mathbb{C}^2$. If h_j are the normalised Hermite functions on $\mathbb{R}$, then we have $(\tilde{\mu}_r(\zeta)h_j, h_k) = 0$ for all j, k. This means that

$$\int_{|z|=r} e^{-i[z,\zeta]} \Phi_{jk}(z)\, d\mu_r = 0$$

where Φ_{jk} are the special Hermite functions.

Using the explicit formulas of Φ_{jk} stated in Proposition 1.4.2, we see that the above equation becomes

$$\int_0^{2\pi} e^{-i[re^{i\theta},\zeta]} e^{im\theta}\, d\theta = 0$$

for all integers m. But this is impossible since the function $\theta \to e^{-i[re^{i\theta},\bar\zeta]}$ is nontrivial. Hence our claim is proved. ∎

The above example shows that it is too much to expect a common zero of the ideal $J(f)$. On the other hand, we have proved in Theorem 4.4.2 that if f is tempered and spherically mean-periodic, then there is a φ_k in $V(f)^\perp$. This means that $\varphi_k \times g^* = 0$ for all $g^* \in V(f)^\perp$. Consequently, $g \times \varphi_k = 0$ or taking the Weyl transform, $W(g)P_k = 0$. Therefore, $W(g)\Phi_\alpha = 0$ whenever $g^* \in V(f)^\perp$ and $|\alpha| = k$. So, Theorem 4.4.2 can be rephrased as follows: if f is spherically mean periodic and is of tempered growth, then

$$\cap\{Ker(\tilde g(0)) : g^* \in V(f)^\perp\}$$

is nonempty. Therefore, it is reasonable to ask the following question even in the general case: does there exist a $\zeta \in \mathbb{C}^{2n}$ such that

$$\cap\{Ker(\tilde g(\zeta)) : g^* \in V(f)^\perp\}$$

is nonempty? By considering the Hermite basis, we may rephrase the question as follows. Let

$$J_\alpha(f) = \{\tilde g(\zeta)\Phi_\alpha : g^* \in V(f)^\perp\}.$$

We are interested in knowing if $J_\alpha(f)$ has a common zero for some α.

Theorem 4.4.7 *Let f be a spherically mean-periodic function on the phase space $\mathbb{C}^n$. Then the following two conditions are equivalent: (i) ζ is a common zero of $J_\alpha(f)$ (ii) $e^{i[z,\bar\zeta]}\Phi_{\alpha,\alpha}(z) \in V(f)$.*

Proof: That (i) implies (ii) is easy to see. If ζ is a common zero of $J_\alpha(f)$, then

$$\int_{\mathbb{C}^n} e^{-i[z,\zeta]}\Phi_{\alpha,\alpha}(z)g(z)\,dz = (\tilde g(\zeta)\Phi_\alpha, \Phi_\alpha) = 0$$

for all $g^* \in V(f)^\perp$. This means that

$$\int_{\mathbb{C}^n} e^{-i[z,\bar\zeta]}\Phi_{\alpha,\alpha}(z)\bar g(z)\,dz = 0$$

and consequently, $e^{i[z,\bar\zeta]}\Phi_{\alpha,\alpha}(z) \in V(f)$.

Conversely, suppose we are given (ii), i.e.,

$$\int_{\mathbb{C}^n} e^{i[z-w,\bar{\zeta}]} \Phi_{\alpha,\alpha}(z-w) e^{\frac{i}{2}Imz.\bar{w}} g^*(w)\, dw = 0$$

for all $g^* \in V(f)^{\perp}$. Recalling the definition of g^*, this means that

$$\int_{\mathbb{C}^n} e^{-i[w,\zeta]} \Phi_{\alpha,\alpha}(z+w) e^{\frac{i}{2}Imz.\bar{w}} g(w)\, dw = 0$$

for all $g^* \in V(f)^{\perp}$. Expanding $\Phi_{\alpha,\alpha}(z+w)e^{\frac{i}{2}Imz.\bar{w}}$ in terms of the special Hermite functions and using their orthogonality properties, we have

$$\Phi_{\alpha,\alpha}(z+w)e^{\frac{i}{2}Imz.\bar{w}}$$

$$= (2\pi)^{\frac{n}{2}} \sum_{\beta} \Phi_{\alpha,\beta}(w)\Phi_{\beta,\alpha}(z).$$

Using this in the last equation we get

$$\sum_{\beta} \Phi_{\beta,\alpha}(z) \int_{\mathbb{C}^n} e^{-i[w,\zeta]} \Phi_{\alpha,\beta}(w) g(w)\, dw$$

$$= \sum_{\beta} \Phi_{\beta,\alpha}(z)(\tilde{g}(\zeta)\Phi_{\alpha}, \Phi_{\beta}) = 0.$$

As $\{\Phi_{\alpha,\beta}\}$ is an orthonormal basis the above is possible only if

$$(\tilde{g}(\zeta)\Phi_{\alpha}, \Phi_{\beta}) = 0$$

for all β which means $\tilde{g}(\zeta)\Phi_{\alpha} = 0$. Since this is true for all $g^* \in V(f)^{\perp}$ we get (i). $\blacksquare$

In the above proof we have not used the invariance of $V(f)$ under the action of $U(n)$, and as such, the theorem remains true in the case of mean-periodic functions as well. In the case of spherically mean-periodic functions, we have the following strengthening of the theorem. Let $J_k(f)$ be the span of $\{J_{\alpha}(f) : |\alpha| = k\}$.

Corollary 4.4.8 *Let f be a spherically mean-periodic function. If ζ is a common zero of $J_k(f)$ and if $\zeta_1^2 + \zeta_2^2 + ... + \zeta_{2n}^2 = a^2$, $a \in \mathbb{C}$, then any $w \in \mathbb{C}^{2n}$ with $w_1^2 + w_2^2 + ... + w_{2n}^2 = a^2$ is also a common zero of $J_k(f)$.*

Proof: If ζ and w are as in the hypothesis, then $w = \sigma\zeta$ for some $\sigma \in U(n)$. Now the identity

$$\tilde{g}(\sigma\zeta) = \mu(\sigma)\tilde{g}_\sigma(\zeta)\mu(\sigma)^*$$

shows that

$$(\tilde{g}(\sigma\zeta)\Phi_\alpha, \Phi_\beta) = (\tilde{g}_\sigma(\zeta)\mu(\sigma)^*\Phi_\alpha, \mu(\sigma)^*\Phi_\beta).$$

Since $\mu(\sigma)$ preserves the eigenspace spanned by $\{\Phi_\alpha : |\alpha| = k\}$, it follows that

$$(\tilde{g}(\sigma\zeta)\Phi_\alpha, \Phi_\beta) = \sum_{|\nu|=k} c_\nu(\tilde{g}_\sigma(\zeta)\Phi_\nu, \mu(\sigma)^*\Phi_\beta).$$

If ζ is a common zero of $J_k(f)$ then the right hand side is zero and hence $\tilde{g}(w)\Phi_\alpha = \tilde{g}(\sigma\zeta)\Phi_\alpha = 0$ for all $g^* \in V(f)^\perp$. Hence the corollary. ∎

The above theorem strongly suggests that functions of the form $e^{i[z,\zeta]}\Phi_{\alpha,\alpha}(z)$, or more generally $e^{i[z,\zeta]}V_\varphi(\varphi, z)$, are the natural counterparts of the exponentials $e^{i\zeta x}$. The analogue of the Schwartz theorem will be the following: if f is a spherically mean-periodic function on the phase space, then for some $\zeta \in \mathbb{C}^{2n}$ and $\varphi \in L^2(\mathbb{R}^n)$, the function $e^{i[z,\zeta]}V_\varphi(\varphi, z)$ belongs to $V(f)$. This conjecture is still open except for the case of tempered functions. General results concerning proper ideals in the space of entire functions of exponential type do not apply here since $J_\alpha(f)$ is not an ideal. Nevertheless, from $J_\alpha(f)$ we can form an ideal to which we can apply known results. But then the problem becomes that of checking whether the associated ideal is proper or not.

Define $I_k(f) \subset \hat{E}(\mathbb{R}^{2n})$ to be the closed ideal generated by

$$\{(\tilde{g}(\zeta)\Phi_\alpha, \Phi_\beta) : g^* \in V(f)^\perp, \alpha, \beta \in N^n, |\alpha| = k\}.$$

Then we have the following result.

Theorem 4.4.9 *Let f be a spherically mean-periodic function on the phase space. Then $V(f)$ contains $e^{i[z,\zeta]}\varphi_k(z)$ if and only if $I_k(f)$ is a proper ideal in $\hat{E}(\mathbb{R}^{2n})$.*

Proof: It is clear from Theorem 4.4.7 that $V(f)$ contains the function $e^{i[z,\zeta]}\varphi_k(z)$ when ζ is a common zero of $I_k(f)$. The converse is also true. To see this, suppose $e^{i[z,\zeta]}\varphi_k(z) \in V(f)$. Proceeding as in the proof of Theorem 4.4.7, and using the fact that

$$\varphi_k(z) = (2\pi)^{\frac{n}{2}} \sum_{|\alpha|=k} \Phi_{\alpha,\alpha}(z),$$

we get the equation

$$\sum_{\beta} \sum_{|\alpha|=k} \Phi_{\beta,\alpha}(z)(\tilde{g}(\zeta)\Phi_\alpha, \Phi_\beta) = 0.$$

As before, this is possible only if $(\tilde{g}(\zeta)\Phi_\alpha, \Phi_\beta) = 0$ for all β and all $\alpha, |\alpha| = k$. But then ζ will be a common zero of the ideal $I_k(f)$.

Therefore, to prove the theorem we have to show that $I_k(f)$ is proper if and only if it has a common zero. To prove this we apply the result of Brown et al [10] which says that any closed ideal in $\hat{E}(\mathbb{R}^{2n})$, invariant under rotations, has a common zero if and only if it is proper. Thus it remains to show that $I_k(f)$ is rotation invariant. But this is again a consequence of the invariance of $J_k(f)$ under the action of the metaplectic representation. If $F(\zeta) \in I_k(f)$ is of the form

$$F(\zeta) = \sum_{|\beta|\leq m} \sum_{|\alpha|=k} c_{\alpha,\beta} h_{\alpha,\beta}(\zeta)(\tilde{g}(\zeta)\Phi_\alpha, \Phi_\beta)$$

then $F(\sigma\zeta)$ is of the same form due to the fact that

$$(\tilde{g}(\sigma\zeta)\Phi_\alpha, \Phi_\beta) = (\tilde{g}_\sigma(\zeta)\mu(\sigma)^*\Phi_\alpha, \mu(\sigma)^*\Phi_\beta)$$

and the k-th eigenspace of the Hermite operator is left invariant by $\mu(\sigma)$. Thus $F(\sigma\zeta)$ is also in $I_k(f)$. This proves the rotation invariance of $I_k(f)$ to which we can apply the result of Brown et al.   ∎

Thus an affirmative answer to our conjecture on mean-periodic functions rests on a positive answer to the following question: if f is a spherically mean-periodic function, then is it true that for some k, the ideal I_k is proper? As $I_k(f)$ is the ideal generated by the subspace $J_k(f)$, it is hard to determine when $I_k(f)$ is proper. Finally, we conclude this section with the following conjecture on mean-periodic functions on the reduced Heisenberg group: if f is a spherically mean-periodic function on the reduced Heisenberg group, then $V(f)$ contains either an exponential function in z or a function of the form $e^{ijt}e^{i[z,\bar{\zeta}]}\varphi_k^j(z)$. To answer this conjecture, we have to prove the corresponding conjecture on mean-periodic functions on phase space!

4.5   Notes and references

The restriction theorem for the reduced Heisenberg group is proved in [81]. There, we have also proved restriction theorems for the cartesian products of Heisenberg groups. Theorem 4.2.2 is taken from Rudin [58] and Theorem 4.2.3 from Rawat [55]. The maximal theorem for spherical means is proved in Nevo-Thangavelu [50] where an ergodic version can also be found. Mean-periodic functions on phase space were studied in [88]. For mean periodic-functions and the two-radius theorem, we refer to Delsarte [16]. For results concerning rotation invariant ideals in $\hat{E}(\mathbb{R}^{2n})$ we refer to Brown et al [10] where Pompeiu transform is dealt with. For more about the Pompeiu problem, we refer to the excellent survey of Bagchi-Sitaram [4].

Bibliography

[1] M. AGRANOVSKY, C. BERENSTEIN, D .C. CHANG and D. PASCUAS, Injectivity of the Pompeiu transform in the Heisenberg group, *J. Anal. Math.*, 63 (1994), 131-173.

[2] S. ANDO, Paley-Wiener type theorem for the Heisenberg group, *Proc. Japan Acad.*, 52 (1976), 331-333.

[3] S. BAGCHI and A. SITARAM, Spherical mean periodic functions on semisimple Lie groups, *Pacific J. Math.*, 84 (1979), 241-250.

[4] S. BAGCHI and A. SITARAM, The Pompeiu problem revisited, *L'Enseignement Math.*, 36 (1990), 67-91.

[5] R. BEALS and P. GREINER, *Calculus on Heisenberg manifolds*, Princeton Univ. Press, Princeton, N. J., (1988).

[6] W. BECKNER, Inequalities in Fourier Analysis, *Ann. Math.*, 102 (1975), 159-182.

[7] C. BENSON, J. JENKINS and G. RATCLIFF, Bounded K-spherical functions on the Heisenberg group, *J. Funct. Anal.*, 105 (1992), 409-443.

[8] G. D. BIRKHOFF, Proof of the ergodic theorem, *Proc. Nat. Acad. Sci., U.S.A.*, 17 (1931), 656-660.

[9] J. BOURGAIN, Averages in the plane over convex curves and maximal operators, *J. Analyse Math.*, 47 (1986), 69-85.

[10] L. BROWN, B. M. SCHREIBER and B. A. TAYLOR, Spectral synthesis and the Pompeiu problem, *Ann. Inst. Fourier, Grenoble*, 23 (1973), 125-154.

[11] L. CARLESON and P. SJÖLIN, Oscillatory integrals and a multiplier problem for the disc, *Studia Math.*, 44 (1972), 287-299.

[12] M. CHRIST, Hilbert transforms along curves: I. Nilpotent groups, *Ann. Math.*, 122 (1985), 575-596.

[13] M. CHRIST, L^p bounds for spectral multipliers on nilpotent groups, *Trans. Amer. Math. Soc.*, 328 (1991), 73-81.

[14] M. COWLING, On Littlewood-Paley-Stein theory, *Suppl. Rendiconti Circ. Mat. Palermo,* 1(1981), 1-20.

[15] Y. C. DAVIS and K. M. CHANG, *Lectures on Bochner-Riesz means*, Cambridge Univ. Press, Cambridge (1987).

[16] J. DELSARTE, *Lectures on topics in mean periodic functions and the two-radius theorem*, Tata Institute of Fundamental Research, Mumbai (1961).

[17] J. DIXMIER, Opérateurs de rang fini dans les représentations unitaires, *Publi. Math. Inst. Htes. Etudes.*, 6(1960), 13-25.

[18] W. F. DONOGHUE, *Distributions and Fourier transforms*, Academic Press, New York, (1969).

[19] H. DYM and H. P. McKEAN, *Fourier series and integrals*, Academic Press, New York (1972).

[20] A. ERDELYI, W. MAGNUS, F. OBERHETTINGER and F. G. TRICOMI, *Higher transcendental functions*, McGraw Hill, New York (1953).

[21] J. FARAUT and K. HARZALLAH, *Deux cours d'analyse harmonique*, Birkhäuser, Boston (1987).

[22] C. FEFFERMAN, The multiplier problem for the ball, *Ann. Math.*, 94 (1971), 330-336.

[23] G. B. FOLLAND, Applications of analysis on nilpotent groups to partial differential equations, *Bull. Amer. Math. Soc.*, 83 (1977), 912-930.

[24] G. B. FOLLAND and J. J. KOHN, *The Neumann problem for the Cauchy-Riemann complex*, Princeton Univ. Press, Princeton, N.J.,(1972).

[25] G. B. FOLLAND, Subelliptic estimates and function spaces on Nilpotent Lie groups, *Ark. Math.*, 13 (1975), 161-207.

[26] G. B. FOLLAND, *Harmonic analysis in phase space,* Ann. Math. Study, 112 (1989).

[27] G. B. FOLLAND, *Introduction to partial differential equations,* Princeton Univ. Press, Princeton, N.J.,(1995).

[28] G. B. FOLLAND and A. SITARAM, The uncertainty principle: a mathematical survey , *J. Fourier Anal. Appl.,* 3 (1997), 207-238.

[29] G. B. FOLLAND and E. M. STEIN, *Hardy spaces on homogeneous groups,* Princeton Univ. Press, Princeton, N.J., (1982).

[30] D. GELLER, Fourier analysis on the Heisenberg group I : Schwartz space, *J. Funct. Anal.,* 36 (1980), 205-254.

[31] D. GELLER, Spherical harmonics, the Weyl transform and the Fourier transform on the Heisenberg group, *Canad. J. Math.* 36 (1984), 615-684.

[32] D. I. GUREVICH, Counter examples to a problem of L. Schwartz, *Funct. Anal. Appl.,* 197 (1975), 116-120.

[33] L. HÖRMANDER, Estimates for translation invariant operators in L^p space, *Acta Math.,* 104 (1960), 93-139.

[34] R. HOWE, On the role of the Heisenberg group in harmonic analysis, *Bull. Amer. Math. Soc.,* 3 (1980), 821-843.

[35] A. HULANICKI and F. RICCI, A Tauberian theorem and tangential convergence for boundary harmonic functions on balls in $\mathbb{C}^n$, *Invent. Math.,* 62 (1980), 325-331.

[36] A. KORANYI, Geometric aspects of analysis on the Heisenberg group, in *Topics in modern harmonic analysis, vol. II,* Proc. Sem. Torino and Milano (1982), 209-258.

[37] R. KUNZE, L^p Fourier transforms on locally compact unimodular groups, *Trans. Amer. Math. Soc.,* 89 (1958), 519-540.

[38] H. LEPTIN, On group algebras of nilpotent Lie groups, *Studia Math.,* 47 (1973), 37-49.

[39] G. MAUCERI, Riesz means for the eigenfunction expansions for a class of hypoelliptic differential operators, *Ann. Inst. Fourier, Grenoble,* 31 (1981), 115-140.

[40] G. MAUCERI, Zonal multipliers on the Heisenberg group, *Pacific J. Math.,* 95 (1981), 143-159.

[41] L. de MICHELE and G. MAUCERI, L^p multipliers on the Heisenberg group, *Michigan Math. J.,* 26 (1979), 361-373.

[42] L. de MICHELE and G. MAUCERI, H^p multipliers on stratified groups, *Ann. Mat. Pura Appl.,* 148 (1987), 353-366.

[43] S. G. MIHLIN, On the multipliers of Fourier integrals (Russian), *Dok. Akad. Nauk.,* 109 (1956), 701-703.

[44] D.MÜLLER, A restriction theorem for the Heisenberg group, *Ann. Math.,* 131 (1990), 567-587.

[45] D. MÜLLER, On Riesz means of eigenfunctions for the Kohn-Laplacian, *J. Reine Angew. Math.,* 401 (1989), 113-121.

[46] D. MÜLLER and E. M. STEIN, On spectral multipliers for Heisenberg and related groups, *J. Math. Pures Appl.,* 73 (1994), 413-440.

[47] R. NARASIMHAN, *Analysis on real and complex manifolds,* North Holland (1985).

[48] A. NEVO and E. STEIN, A generalisation of Birkhoff's pointwise ergodic theorem, *Acta Math.,* 173 (1994), 135-154.

[49] A. NEVO and E. STEIN, Analogues of Wiener's Ergodic theorems for semisimple groups, *Ann. Math.,* 145 (1997), 565-595.

[50] A. NEVO and S. THANGAVELU, Pointwise ergodic theorems for radial averages on the Heisenberg group, *Advances in Math.,* 127 (1997), 307-334.

[51] J. PEETRE, The Weyl transform and Laguerre polynomials, *Matematiche (Catania),* 27 (1972), 301-323.

[52] J. PEETRE and G. SPARR, Interpolation and noncommutative integration, *Ann. Mat. Pura Appl.,* CIV (1975),187-207.

[53] D. L. RAGOZIN, Central measures on compact simple Lie groups, *J. Funct. Anal.*, 10 (1972), 212-229.

[54] P. K. RATNAKUMAR, R. RAWAT and S. THANGAVELU, A restriction theorem for the Heisenberg motion group, *Studia Math.*, (to appear)

[55] R. RAWAT, A theorem of Wiener-Tauberian type for $L^1(H^n)$, *Proc. Indian Acad. Sci.*, (in press)

[56] R. RAWAT and A. SITARAM, The injectivity of the Pompeiu transform and L^p analogues of Wiener-Tauberian theorem, *Israel J. Math.*, 91 (1995), 307-316.

[57] M. REED and B. SIMON, *Methods of modern mathematical physics I: Functional Analysis*, Academic Press, New York (1972).

[58] W. RUDIN, *Functional analysis*, Mc Graw-Hill, New York (1973).

[59] L. SCHWARTZ, Theorie generale des functions moyenne-periodique, *Ann. Math.*, 48 (1947), 857-928.

[60] A. SITARAM, M. SUNDARI and S. THANGAVELU, Uncertainty principles on certain Lie groups, *Proc. Indian Acad. Sci.*, 105 (1995), 135-151.

[61] C. D. SOGGE, Oscillatory integrals and spherical harmonics, *Duke Math. J.*, 53 (1986), 43-65.

[62] C. D. SOGGE, *Fourier integrals in classical analysis*, Cambridge Univ. Press, Cambridge , (1993).

[63] E. M. STEIN, *Topics in harmonic analysis related to Littlewood-Paley theory*, Ann. Math. Study., 63 (1971).

[64] E. M. STEIN, Maximal functions: spherical means, *Proc. Natl. Acad. Sci. U.S.A*, 73 (1976), 2174 - 2175.

[65] E. M. STEIN, *Harmonic Analysis*, Princeton Univ. press, Princeton, (1993).

[66] E. M. STEIN and S. WAINGER, Problems in harmonic analysis related to curvature, *Bull. Amer. Math. Soc.*, 84 (1978), 1239 - 1295.

[67] K. STEMPAK, On convolution products of radial measures on the Heisenberg group, *Colloq. Math.,* 50 (1985), 125-128.

[68] R. S. STRICHARTZ, Multipliers for spherical harmonic expansions, *Trans. Amer. Math. Soc.,* 167 (1972), 115-124.

[69] R. S. STRICHARTZ, Sub-Riemannian geometry, *J. Differential Geometry,* 24 (1986), 221-263, Correction 30 (1989), 595-596.

[70] R. S. STRICHARTZ, Harmonic analysis as spectral theory of Laplacians, *J. Funct.Anal.,* 87 (1989), 51-148.

[71] R. S. STRICHARTZ, L^p harmonic analysis and Radon transform on the Heisenberg group, *J. Funct. Anal.,* 96 (1991), 350-406.

[72] G. SZEGÖ, *Orthogonal polynomials,* Amer. Math. Soc., Colloq. publ., Providence, RI,(1967).

[73] M. E. TAYLOR, *Non commutative harmonic analysis,* Amer. Math. Soc., Providence, RI, (1986).

[74] M. E. TAYLOR, *Non commutative microlocal analysis, Part I,* Memoirs Amer. Math. Soc., (1994).

[75] F. TREVES, *Topological vector spaces, distributions and kernels,* Academic Press, New York (1967).

[76] S. THANGAVELU, Multipliers for Hermite expansions, *Revist. Math. Ibero.,* 3 (1987), 1-24.

[77] S. THANGAVELU, Littlewood-Paley-Stein theory on $\mathbb{C}^n$ and Weyl multipliers, *Revist. Math. Ibero.,* 6 (1990), 75-90.

[78] S. THANGAVELU, Riesz means for the sublaplacian on the Heisenberg group, *Proc. Indian Acad. Sci.,* 100 (1990), 147-156.

[79] S. THANGAVELU, A multiplier theorem for the sublaplacian on the Heisenberg group, *Proc. Indian Acad. Sci.,* 101 (1991), 169-177.

[80] S. THANGAVELU, Restriction theorems for the Heisenberg group,*J. Reine Angew. Math.,* 414 (1991), 51-65.

[81] S. THANGAVELU, Some restriction theorems for the Heisenberg group, *Stud. Math.,* 99 (1991), 11-21.

[82] S. THANGAVELU, Spherical means on the Heisenberg group and a restriction theorem for the symplectic Fourier transform, *Revist. Mat. Ibero.* 7 (1991), 135-155.

[83] S. THANGAVELU, On regularity of twisted spherical means and special Hermite expansions, *Proc. Ind. Acad. of Sci.,* 103, (1993), 303-320.

[84] S. THANGAVELU, *Lectures on Hermite and Laguerre expansions,* Mathematical notes, 42, Princeton Univ. press, Princeton, (1993).

[85] S. THANGAVELU, On Paley-Wiener theorems for the Heisenberg group, *J. Funct. Anal.,* 115 (1993), 24-44.

[86] S. THANGAVELU, A Paley-Wiener theorem for step two nilpotent Lie groups, *Revist. Mat. Ibero.,* 10 (1994), 177-187.

[87] S. THANGAVELU, Spherical means and CR functions on the Heisenberg group, *J. Anal. Math.,* 63 (1994), 255-286.

[88] S. THANGAVELU, Mean periodic functions on phase space and the Pompeiu problem with a twist, *Ann. Inst. Fourier, Grenoble,* 45 (1995), 1007-1035.

[89] S. THANGAVELU, On Paley-Wiener properties of nilpotent Lie groups, (unpublished manuscript) (1996).

[90] G. N. WATSON, *A treatise on the theory of Bessel functions,* Cambridge Univ. Press, London (1966).

[91] Y. WEIT, On Schwartz theorem for the motion group, *Ann. Inst. Fourier, Grenoble,* 30 (1980), 91-107.

[92] N. WIENER, *The Fourier integral and certain of its applications,* Cambridge Univ. Press, London (1933).

[93] A. ZYGMUND, On Fourier coefficients and transforms of functions of two varibles, *Studia Math.,* 50 (1974), 189-202.

Index

Progress in Mathematics

Edited by:

Hyman Bass
Dept. of Mathematics
Columbia University
New York, NY 10010
USA

J. Oesterlé
Institut Henri Poincaré
11, rue Pierre et Marie Curie
75231 Paris Cedex 05
FRANCE

A. Weinstein
Department of Mathematics
University of California
Berkeley, CA 94720
USA

Progress in Mathematics is a series of books intended for professional mathematicians and scientists, encompassing all areas of pure mathematics. This distinguished series, which began in 1979, includes authored monographs and edited collections of papers on important research developments as well as expositions of particular subject areas.

We encourage preparation of manuscripts in some form of TEX for delivery in camera-ready copy which leads to rapid publication, or in electronic form for interfacing with laser printers or typesetters.

Proposals should be sent directly to the editors or to: Birkhäuser Boston, 675 Massachusetts Avenue, Cambridge, MA 02139, U. S. A.